crossing over

Other Books by the Authors

Illuminations

Finders, Keepers

Also by Rosamond W. Purcell

A Matter of Time

Half Life

A Collector's Catalogue

Special Cases

PHOTOGRAPHS FOR:

Suspended Animation

Curiositeit

Swift as a Shadow

Also by Stephen Jay Gould

Ontogeny and Phylogeny

Ever Since Darwin

The Panda's Thumb

The Mismeasure of Man

Hen's Teeth and Horse's Toes

The Flamingo's Smile

An Urchin in the Storm

Time's Arrow, Time's Cycle

Wonderful Life

Bully for Brontosaurus

Eight Little Piggies

Dinosaur in a Haystack

Full House

Questioning the Millennium

Leonardo's Mountain of Clams and the Diet of Worms

Rocks of Ages

The Lying Stones of Marrakech

crossing over

where art and science meet

Stephen Jay Gould

and

Rosamond Wolff Purcell

THREE RIVERS PRESS • NEW YORK

Published by Three Rivers Press, New York, New York.
Member of the Crown Publishing Group.

Random House, Inc. New York, Toronto, London, Sydney, Auckland
www. randomhouse.com

Three Rivers Press is a registered trademark and the Three Rivers Press colophon is a trademark of Random House, Inc.

All the essays contained in this work were previously published in *Harvard Magazine* and *The Sciences*.

Printed in Hong Kong

DESIGN BY LYNNE AMFT

Library of Congress Cataloging-in-Publication Data
Gould, Stephen Jay.
Crossing over / by Stephen Jay Gould; photographs by Rosamond Purcell. — 1st ed.
1. Science. I. Purcell, Rosamond Wolff. II. Title.
Q171.G685 2000
500—dc21 99-05834

ISBN 0-609-80586-X

10 9 8 7 6 5 4 3 2 1

First Edition

For Richard Milner

—SJG

For Jim Mairs

—RWP

contents

II Mind and Nature

preface
crossing the bar

Most famous quotations are misattributed, "improved," or downright fabricated. (As Yogi Berra, America's most celebrated public philosopher, may have stated: "I didn't say half the things I said.") On the subject of texts and images, Confucius almost surely did not utter his most famous pearl of wisdom: "One picture is worth a thousand words." (Some versions reset the equation to ten thousand words; I only average seven hundred or so in the essays of this book.) In any case, the message persists, even though the true originator disappeared long ago into the mists of antiquity or prehistory. (I will vote for Cro-Magnon Charlie, the great bison painter of the oldest Paleolithic cave of Chauvet, as an unrecorded initial source for this quintessential truism.)

Nonetheless, while affirming the general sentiment behind this pseudo-Confucian wisdom, I must reject the literal claim in one vital respect, at least for the conjunction of words and images in this book, and probably for the junction of pictures and explanations in general. If a text merely tries to describe an existing picture, then the conventional equation may hold. But often, as in this work, pictures and words represent two modalities for equal and conjoined reflection. Thus, my text does not provide an explanation or explication of Rosamond Purcell's photographs, but rather a musing (perhaps, on occasion, amusing as well) upon the common theme of a truly joint effort.

We have produced two previous books in this mode (*Illuminations* and *Finders, Keepers*), although we regard this third effort as a Goldilockean golden mean between a text that may have been too cryptic and sparing in the first, and too generous in the second. We developed this format by prolonged trial and error, by first publishing the collaborations of this book in two magazine series: *Fossil Endpapers* for *Harvard Magazine*, and *On Common Ground*, still continuing in *The Sciences*.

Perhaps we have finally achieved, by distillation, not only a better balance but, more important, an integration that might truly dissolve the usual contrast (or even any meaningful distinction) between word and image. We record this goal in our book's title, *Crossing Over*, with its immediately literal aim of good conversation between art and science, photography and text—but also in its familiar metaphorical meaning of attempts to reach a better and higher place. We may further explicate these intentions by reference to a particular essay in this book, entitled "Part, Counterpart." As depicted therein, most fossils break into two sections when a collector or scientist splits a rock to reveal the specimen on a bedding plane: the part and the counterpart.

Admittedly, in some fossils, these two sections cannot claim equality—for, in solid and discrete objects like shells or bones, we call the object itself the "part" and its impression into the overlying rock the "counterpart." Actual entities trump imprints (thumb vs. thumbprint) in our usual assessments of interest and worth. But, in complex fossils made of many small pieces or segments (a crinoid composed of thousands of ossicles, or an arthropod with numerous jointed appendages), the enclosing rock never splits "perfectly," with all substance going to one side and all impression to the other. Each section will contain an amalgam of positive parts and negative imprints—and the designation of one side as the "part" and the other as the "counterpart" becomes merely conventional.

We view the pieces in this book as parts and counterparts of such complex totalities. Neither the picture nor the text holds much meaning without its complement; neither is more important than the other; neither simply explicates the other. The pictures and text build (at least in our hopes and intentions) a unified and inextricable totality. As another symbol of this attempt, consider the literal linkage of part and counterpart in the photograph on page 31. The "rope" may only be an adventitious crack, filled with a lovely chain of manganese oxide dendrites, but this chain hangs like the true catenary of a suspension bridge—representing both a concrete and a metaphorical linkage between the two equal and complementary halves of each essay.

Small minds have usually viewed Science and Art as adversarial—at least from Goethe's complaints about narrow-minded naturalists who would not take his anatomical and geological works seriously because he maintained a day job as a poet, to C. P. Snow's identification and lament about two noncommunicating cultures in the most widely discussed example of our generation. But the unifying modes and themes of human creativity surely transcend the admitted differences of subject matter in these two realms of greatest interest and occasional (even frequent) triumph of both heart and mind. We see nothing anomalous, or even the slightest bit strange, about this double integration of science with art, and text with image.

The trade secret of the literary essayist fits the photographs of this book particularly well: Love, treasure, and exemplify the little details in all their intricate fascination, but not for themselves alone (or you will be buried by the two adjectival curses of academic failure and narrowness: *arcane* and *vacuous*). Instead, use those lovely details to exemplify, in an explicit mote of concrete that can carry a general message with light and lightness, the great issues of matter and meaning that only a pompous fool would attack head on. (No one should ever write a tendentious and general essay about the meaning of human distinctiveness, but a picture of Fred Astaire next to an ape in similar pose conveys both the challenge and the reality with accessible definition, and maybe even with requisite modesty.)

Thus, although each of these essays features some odd little corner of the natural world (or of human perception and intrusion thereupon), I would venture to state (grandiose as the claim may sound) that these essays treat the two great themes of philosophy through the ages: ontology, or the nature of reality, and epistemology, or how the human mind obtains its knowledge of reality. We have even dared to back up this claim by dividing the essays into two groups, the first ontological, and the second epistemological.

The ontological set includes our musings on the richness and complexities that make our world so fascinating and often so difficult to comprehend (but so unifying if we succeed). We group these little pieces under the four subthemes of dimensionality (focusing on added sources of knowledge beyond a usual and unitary appearance), scaling (following Julian Huxley's maxim that "size has a fascination all its own"), symmetries (in and out, right and left), and loss or gain of information by attention to the layering within apparently discrete objects.

The epistemological second half includes three major themes and hang-ups in the interaction of mind and nature. We first treat some of the mental biases (probably universal traits of human nature, rather than mere cover-ups of particular cultures) that both facilitate our detection of genuine signals within nature's noisy presentation, but that also lead us astray when we mistake our brain's preferences for nature's objectivity—as in our propensity for seeing faces, or our attraction to "properly" symmetrical bodies with all parts present and in the "right" places.

The second subsection discusses taxonomy, or classification in general, as a basic need and imposition of mind upon an otherwise jumbled richness of nature—and not as a boring activity of placing notes in proper pigeonholes or of pasting stamps in the preassigned spaces of an album. This need (as depicted in essay 19, for example) transcends location, profession, or intention—as properly classified mollusc shells build the columns of America's most famous twentieth-century folk sculpture, and also fill the collection of an eighteenth-century Italian museum dedicated to displaying the effects of Noah's flood.

Finally, the last part treats our propensity for ordering nature by seeking patterns, and then telling explanatory stories based on limited themes that express more about human hope and conceptions of valor than about any causal process operating "out there" in nature (for many patterns only record the inevitable clumping generated by truly random systems). The first two essays explore our preferences for tales about progress or development, and the strictly limited ways that our minds permit such tales to "go." The next three pieces (essays 26–28) tell stories about decline and extinction, although we manage to find encouragement with the general message by focusing on limited aspects that we can read in a more encouraging light. Finally, the last two pieces form a contrasting pair about the infiltration of disorder (the penultimate essay 29) into our artificial forms and symmetries, and the reversal of universal deterioration by ingenious, if adventitious, recycling to other equally valid uses for different creatures (in essay 30). In this conversion, without loss of substance or order, of a human book into a rat's nest, perhaps we may find a humble symbol (also a true example, however modest) of the greatest victory that biology can exert over the cosmos: the promise of continuity as facilitated by active wit, rather than passive embedding within nature's unvarying ways.

STEPHEN JAY GOULD

I

form and time

dimension and distinction

inside dimensions

THE SKIN'S OPACITY STILL BUILDS ONE OF THE GREAT BARriers in medicine. Often, the source of illness lies hidden just an inch or two under the body's external cover, quite ripe for cure or removal, if only we could visualize and identify the trouble. For this reason x-radiography quickly became a vital industry, and much of the fanciest and most expensive of modern medical equipment has, as its raison d'être, the more efficient "penetration" of skin to produce an image of internal structures without cutting.

Scholars, teachers, and collectors also strive to understand and represent the insides of bodies by a variety of techniques, including two-dimensional methods of illustration and three-dimensional preservation of dissected body parts. Here we juxtapose these two great methods, one atop the other.

The large figure of a dissected woman, forming the background of this photograph, belongs to a famous set published in 1745 by Jacques-Fabien Gautier d'Agoty, onetime assistant to Jakob Christof Le Blon. Using Isaac Newton's theories, Le Blon had developed a process for producing color prints by superimposing three mezzotint plates inked in Newton's

The Skull in the Belly

three primary colors—yellow, red, and blue—one above the other. He later discovered that he needed a fourth, black plate to blend the colors properly, thereby inventing the four-color printing process still used today.

Gautier d'Agoty soon left Le Blon's employ to set up a rival workshop. Since Le Blon preferred to follow Newton's three-color theory strictly, and therefore remained reluctant to use the additional black plate of his own discovery, Gautier d'Agoty waded in and claimed that he had invented four-color printing himself. Gautier d'Agoty produced large, magnificent anatomical plates more celebrated for their beauty than for their accuracy. One commentator remarked that Gautier d'Agoty's plates "impress the critical observer with their arrogance and charlatanry and do not recommend themselves to the student of anatomy."

The skull belongs to the extensive collection of Joseph Hyrtl of Vienna (1811–1894), purchased in 1874 by the Mütter Museum of the College of Physicians of Philadelphia. Hyrtl may have been a fine anatomist and a skilled preparator, but his claims and methods must be questioned. He told his Philadelphia purchasers, for example, that he owned the skull of Mozart, though no one knows the site of the great composer's burial for certain. (Hyrtl's father played the oboe under Haydn in Count Esterhazy's band, so perhaps this family background contributed to the younger Hyrtl's fantasy or naïveté.) He obtained most of his 139 skulls by theft. The depicted skull bears the label "From the catacombs of St. Stephan," but the catalog entry adds: "Stolen by myself. Skull of a Canonicus."

We make a small effort here to redress the exaggeration of both contributors by depriving each of one dimension, while adding some elegance in the photographic juxtaposition. We reduce Hyrtl to two dimensions by the art of photography, and we have deprived Gautier d'Agoty of his celebrated color. At life size, the background figure is so large that Hyrtl's skull fits into the belly like an entire fetus in the womb. The union seems so complete that I can almost imagine the skull untimely ripped from the poor woman, thus exposing her interior as though nothing in her life became her like this leaving of it.

individuality

THE AESTHETIC AND ETHICAL FOUNDATION OF MODERN Western culture rests firmly on our belief in the distinctiveness of each individual (although many traditions today—and several earlier Western philosophies—do not accept or cherish such a centerpiece). At the climax of Ibsen's great play, *A Doll's House*, Torvald importunes Nora not to leave by reminding her that she is "first and foremost . . . a wife and mother." Nora replies: "I believe that first I am an individual."

But the primacy of individual distinction transcends the preferences of any particular culture. The uniqueness of each individual, and the consequent variability among individuals in biological populations of sexually reproducing organisms, provides the sine qua non for evolution by Darwin's mechanism of natural selection. This primary cause of evolutionary change operates in a paradoxical manner. Natural selection can create nothing

by itself; Darwin's process works by selective elimination and preservation—that is, by imparting higher reproductive success to a subset of individuals fortuitously better adapted to changing local environments. Natural selection can only be effective if the individuals of a population vary extensively, with each different from the others—thus providing enough "raw material" to fuel the process of natural selection.

In the lifestyle of vertebrates, sex and reproduction became intrinsically intertwined—and we often err in assuming that they represent the same biological process, with the same evolutionary significance. But sex and reproduction play markedly different biological roles. Reproduction continues a lineage by making more members; sex provides variability among individuals by mixing the genetic products of two parents in each offspring. Asexual reproduction is far more rapid and efficient (think of a splitting bacterium or a budding hydra) than the *Sturm*, *Drang*, and prolonged gestation of most sexual systems. But asexual reproduction also marks an evolutionary dead end because the numerous offspring of a single parent form a clone with no genetic variability among individual members unless new and rare mutations arise.

Most people may not have encountered this argument about the biological meaning of individuality, but we do know the pattern in a visceral way that defines our intrinsic concept of "normality." That is, we know that each fellow human being should be distinctly different from all others—and we become, to use the modern vernacular, "freaked" or "weirded out" when we come face-to-face with a human clone.

Eng and Chang Bunker, born in Siam (now Thailand) in 1811, gave the name of their birthplace to the general phenomenon of conjoined, or Siamese twinning—the greatest of all challenges to our concept of individuality. Eng and Chang performed for P. T. Barnum, made a pile of money, married a pair of sisters in North Carolina, and fathered about ten children each. (Their domestic arrangements were, needless to say, unconventional: the wives lived in different houses, while Eng and Chang spent half the week in each domicile.) But Eng and Chang, though a cloned pair of identical twins, remained complete and distinct individuals, joined at the abdomen only by a broad band of flesh.

Other styles of Siamese twinning threaten our traditional notion of individuality more directly. Ritta and Christina, born in Sardinia

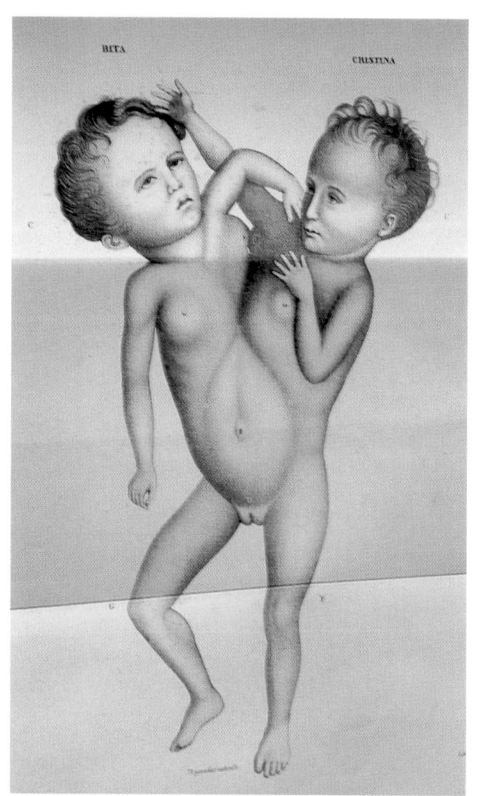

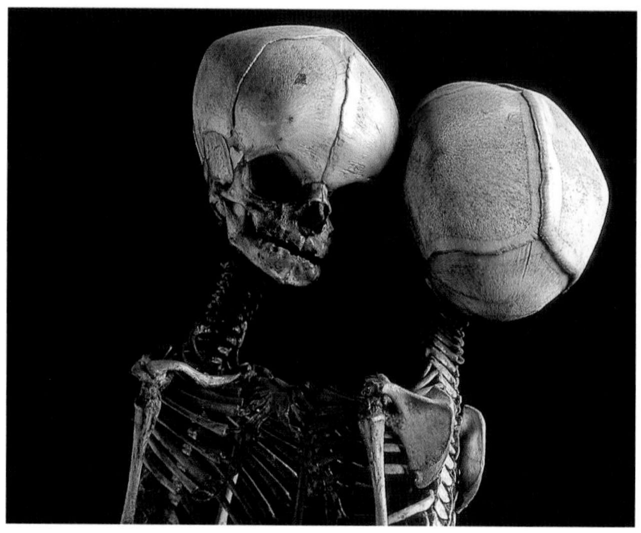

Twins joined at the rib cage

Ritta/Christina
(from 1833 book by Etienne Serres)

Plaster cast of Eng and Chang with reflection

in 1829, had two heads and four arms, but only a single lower body with two legs. When they (or she) died in infancy, probably of overexposure from being exhibited too often (as the impoverished parents tried to make some money from their misfortune), a Paris newspaper raised the question of the day: "Already it is a matter of grave consideration with the spiritualists, whether they had two souls or one." Most commentators invoked an old Western prejudice and opted for two, on the argument that two complete heads carried two brains and, by implication, two souls. But what, then, could be said of the same phenomenon zipped the other way—that is, a conjoined pair with two lower bodies and only one head?

We regard cloning, or the production of two (or more) genetically identical creatures, as eerie beyond all concepts of natural order, at least for mammals and other complex animals. Dolly the sheep, the first mammal cloned from an adult cell of a single parent (and the most famous invocation of her species since John the Baptist designated Jesus as the Lamb of God), shocked the world beyond any merely intellectual reason—primarily by raising for so many people (or so I infer by listening to numerous talk shows and gauging the concerns expressed) our deepest worries about the distinctness of our personhood: Are clones distinct individuals? Does each member of a clone have a soul? Am I still a unique individual if I clone a daughter from a cell scraped from the inner lining of my cheek?

May I suggest, following the preceding discussion, that all these fears are misplaced, for these questions have a clear answer, known to all human societies throughout history. Identical twins are clones—in fact, far closer clones than Dolly and her mother (for identical twins share mitochondrial as well as nuclear DNA; they also gestate in the same womb and usually grow up in the same environment). Yet we know, and have always known, that human identical twins—whatever their quirky similarities in behavioral details as well as physical appearance—become utterly distinct individuals because the unique and contingent pathways of each complex life impose a separate and formative influence upon any individual, even upon the members of a genetically Xeroxed pair. Eng and Chang, closer perforce than any ordinary pair of identical twins, developed distinctly different personalities. One became morose and alcoholic; the other remained benignly cheerful and teetotaling.

Dolly's birth raises legitimate fears—as any powerful new technology must, for all innovations of this magnitude carry potential for fruitful employment or unspeakable misuse. We can all spell out unacceptable scenarios for human cloning, and we should pursue our ethical debates on this subject with vigor and vigilance. Some nonhuman uses are benign and already in practice. (We have been cloning fruit trees by grafting for decades.) Other potential applications grab my fancy. (I would be powerfully intrigued by a horse race among ten identical Citations—what a test for the skills of jockeys and trainers!)

Dolly should inspire our interest and our watchfulness, not our loathing, disgust, or heedless rejection. Human clones remain unambiguous individuals, as proven by the identical twins (or more, up to the Dionne quintuplets and beyond) of all human populations. Environment will impose a cantankerous uniqueness upon any individual. But Darwinian biology—and, therefore, the world of our fathers, our mothers, and all our history—requires the variability conferred by genetic distinctiveness, and therefore defines the domain of our usual experience and visceral comfort.

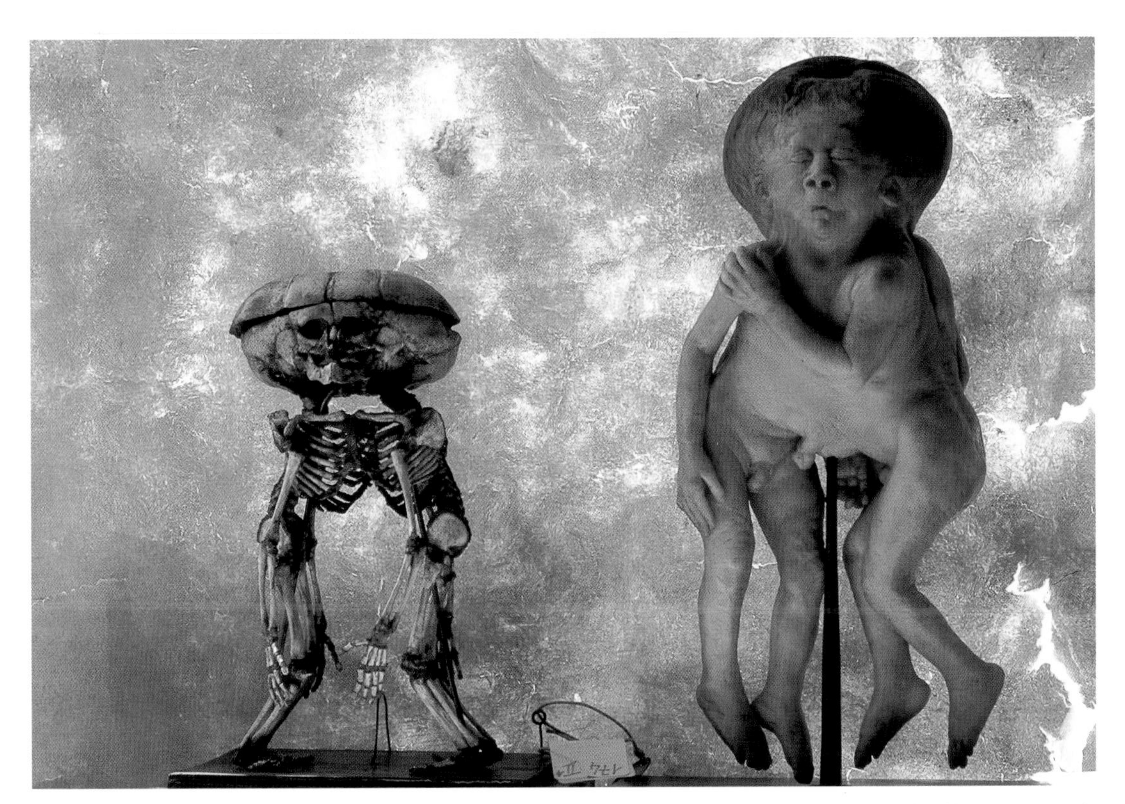

Wax model of conjoined twins with their skeletons (top of skull missing)

Two-headed sheep

3

part, counterpart

When we split a rock to reveal a fossil lying on a bedding plane, we usually get two for the price of one. The fossil itself may adhere to one slab, but its impression (often exquisitely detailed) will be left on the other piece as a mold. Geologists, shunning their usual thirst for jargon and adopting vernacular English for once, call the main slab a "part" and the overlying block carrying the fossil's impression a "counterpart." Part, counterpart; thumb and thumbprint.

We see here both the part and the counterpart of a beautifully preserved fossil from the lithographic limestones of Solnhofen. Since part and counterpart form mirror images, we note a pleasing symmetry, enhanced by a similar bilateral (mirror) symmetry of right and left halves *within* each part (as in the human body). Are we seeing a dexterous Mr. Peanut, nimbly bowing with crossed legs to his sinister counterpart? Or is he a Thin Man of Mardi Gras, wearing a crown and precariously balanced on stilts? He seems to sport a face, with eyes, nose, and smile—so I will opt for Mr. Peanut. But his human form is an illusion. We are really looking at a fossil, shrimplike creature standing

Mr. Peanut

on its head. Mr. Peanut's legs are the shrimp's antennae, his head the fossil's abdomen, and his crown its tail flaps. The best clue to reality can be found in Mr. Peanut's several pairs of arms (the shrimp's legs) —useful, no doubt, if we could evolve such "extras," but as precluded for vertebrates as they became canonical for arthropods.

Another set of specimens might be taken for three panels of an Assyrian frieze, a segment, perhaps, of a parade on their own Parthenon. The figures walk, bow in deference, and pirouette; their domed caps proclaim their ethnicity. This whimsical misperception in human terms also teaches a lesson about biological unity. The internal skeletons of vertebrates and the external carapaces of crustaceans evolved independently—the common ancestor of lobsters and lumberjacks probably possessed no hard parts at all. Yet basic similarities of design evolve again and again because only certain forms and symmetries can be sculpted from organic matter or can perform the basic functions of life.

Here, we think we see something familiar in an inverted shrimp because arthropods and vertebrates share two basic features of complex organic design: bilateral symmetry and articulated skeletons made of rigid pieces connected at joints. If evolutionary solutions knew no limitations, two groups that evolved their skeletal complexity in complete genealogical independence would display no such anchor of common design.

The universe, physicists tell us, exhibits few properties that can impart a direction to time or any preferred asymmetry to matter. Since we need direction to make sense of the world, we impose these concepts of order upon an indifferent nature. A favored technique for ordering imposes a judgment of better or worse, proper or improper, upon two equally probable and worthy ways—for differences in status must impart both direction and asymmetry to any pair of items. Thus, an object may move around the perimeter of a circle in either direction, but we call one route clockwise and the other counterclockwise. A carpenter's screw or a snail's shell may coil in either direction and work just as well—but we call the more frequent state *dextral,* or right-handed, for no defensible reason. So too for part and counterpart. We often view only the part as the "real" fossil, though all pro-

Three specimens of Meccochinus longimanarus

fessionals know that both slabs can yield different and complementary information.

Perhaps this attribution of differential worth, inherent in our names of part and counterpart, can be justified when the part is an entire clamshell, for example, and the counterpart only an impression. But, in these exquisite Solnhofen arthropods, composed of so many separate segments and with soft anatomy preserved as well, fossils rarely break so cleanly that all the goodies go to one side, with only impressions remaining upon the other. In such complex fossils as Mr. Peanut, each slab contains important bits of the organism, and we must study part and counterpart together—for both slabs provide vital information. In fact, we must often resort to an arbitrary convention if we wish to use the traditional terms at all. We call the slab a part when we look down on the back of the animal, and a counterpart when we look up upon the belly.

In any case, the bond and equality of part and counterpart gain a symbolic affirmation for this fossil in the apparent chain or ligature joining part and counterpart at Mr. Peanut's forehead—actually just a crack in the rock festooned with dendrites of manganese oxide. *Ex duobus unum.*

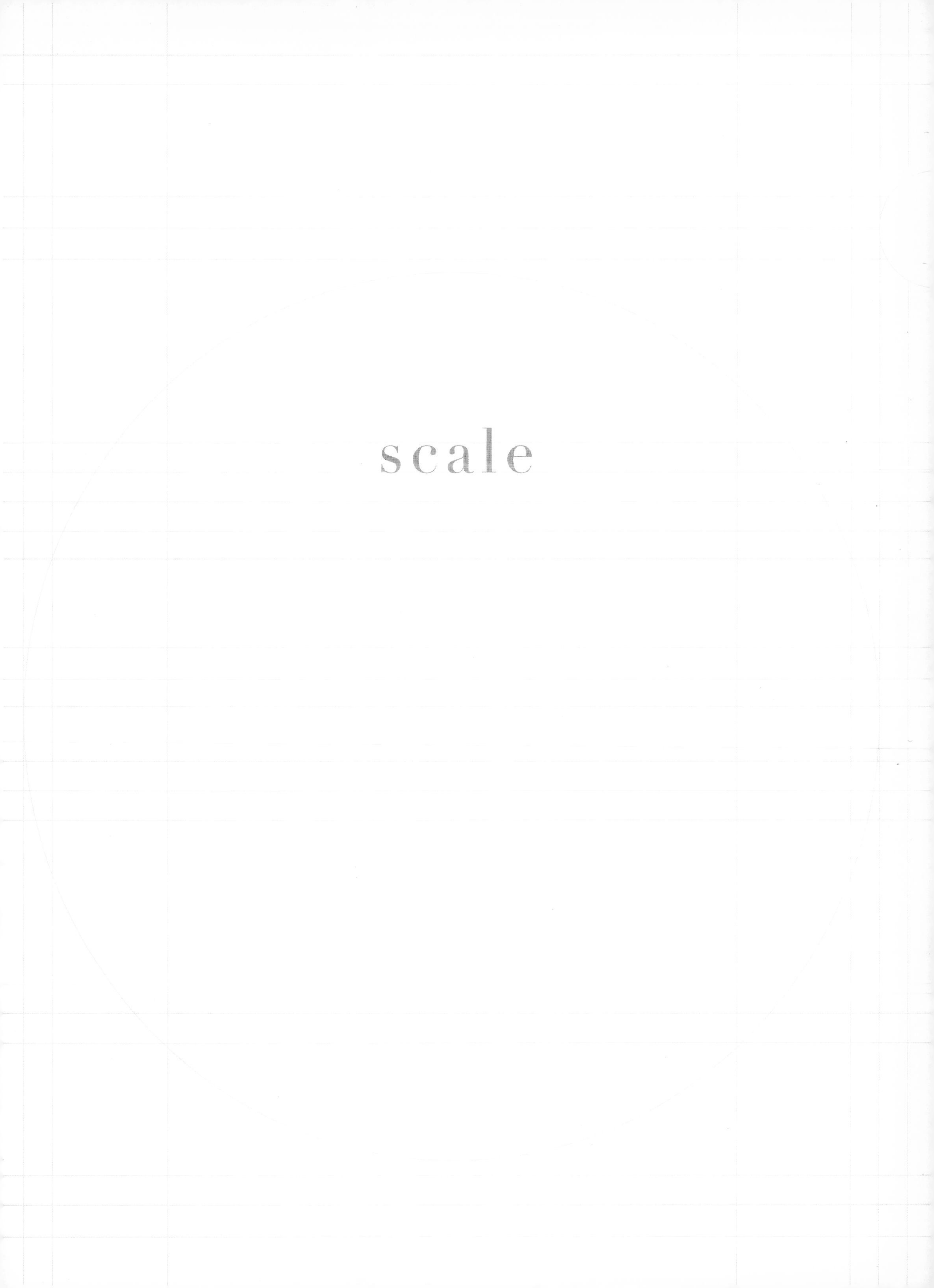

scale

Cut and polished ammonite on piece of jade

scales of destruction

THE CRITERIA OF DEATH OFTEN INCLUDE EXTREME CAPRI-ciousness. Jephthah identified his Ephraimite enemies for slaughter by their inability to pronounce the *sh* in *shibboleth* when they tried to pass for Gileadite allies. John Milton spoke of "that sore battle when so many dy'd":

Without reprieve, adjudged to death
For want of well pronouncing Shibboleth.

The beautiful fossil ammonite, photographed here against a slab of jade, embodies, at grandest scale, this theme of death's almost arbitrary finality. These mollusks became prominent victims in three of the five major mass extinctions that punctuate the earth's fossil record. They almost succumbed in the granddaddy of all great dyings—the Permian event (250 million years ago) that may have eliminated more than 95 percent of marine invertebrate species. They barely persisted through the late Triassic mass extinction just one geological period later. Yet, in each case, one or two lineages sneaked through, and the group rediversified. In the next great wipeout, the

Cretaceous-Tertiary extinction 65 million years ago, not even one lineage survived, and the entire group disappeared forever, along with the dinosaurs and some 50 percent of marine invertebrate species.

When paleontologists held a more deterministic and gradualist view of mass extinctions, they tried to interpret such deaths as failures of adaptation—as extinctions marking a sensible end to superannuated groups that succumbed in competition to more advanced forms at stringent times, when the going got particularly tough. But we now recognize mass extinctions as more frequent, more profound, more rapid, and more unusual in their effects than we previously imagined. The Cretaceous-Tertiary dying, in particular, was almost surely triggered by the impact of a large extraterrestrial body. At such times of truly global catastrophe, the rules change profoundly, and environments alter beyond any power of ordinary adaptation to adjust. Groups live or die for reasons as capricious as the ability to pronounce a meaningless sound. Anatomies that evolved for lifestyles of normal times are either fortuitously suited for passage through the unpredictable catastrophe—or they are not. The ammonites squeaked through twice, but then succumbed to strike three.

Jephthah fought his earlier battles against a people called the Ammonites (though the fossils owe their name to a different Egyptian god of the same title). To secure God's aid in defeating the Ammonites, Jephthah rashly vowed to sacrifice in flames the first living thing to greet him upon his return. I suppose that he expected to encounter a dog or a goat, but his daughter came forth "with timbrels and with dances." In this patriarchal document, Jephthah's tragedy does not reside so much in his daughter's death as in the extinction of his family line, for he has no other child—a theme underscored by her last request to her father for a short stay before the inevitable flames: "Let me alone two months, that I may go up and down upon the mountains, and bewail my virginity." Again, the capricious fires of catastrophe can trigger extinctions at any scale.

Our fractal world expresses self-similarity at all magnitudes, and the photograph on the next page displays some relics of ruin in smaller fires, just as the ammonite represents an artifact of global death. We note a nail from the burned-out visitor's center at Hawaii

Dante's Inferno *with burned lightbulb*

Volcanoes National Park, following burial in a lava flow; and a lightbulb (seemingly inscribed with a delicate Japanese landscape) from a building destroyed in the California canyon fires of 1993. Both artifacts lie on the pages of a burned book—which happens, ever so appropriately, to be a copy of Dante's *Inferno*.

Destruction comes like a thief in the night. But destruction can wipe a slate clean and create space for novelty that would otherwise never have won an opportunity. The pathways may be peculiar and unpredictable at all scales, but the results can be wondrous. London and Chicago rose from the ashes of architectural holocaust. Mammals rose to prominence, and humans can read this book today, because dominant dinosaurs, after prevailing against mammals for more than a hundred million years, succumbed to externally triggered catastrophe. Must we always experience a chariot of fire before we can build Jerusalem on a green and pleasant land?

5

worldly skulls

An entire human lifetime passes in a geological eye blink. The ten thousand years of our recorded civilization, with all its transformations from atlatl to atom bomb, constitute but 1 percent of the million years that our earliest ancestor, *Australopithecus afarensis* (a.k.a. Lucy), lived without any measurable change in Africa.

The scaling of size presents just as fascinating a subject as these differences in realms of time. Scale becomes most elusive in self-similar fractal objects, where a part, blown up, scarcely differs from the whole. A fern leaflet with serrations around the edge looks like the central stem with leaflets around its periphery; the coastline of North America from Maine to Florida may resemble the pathway around a single headland on a rocky shore.

Usually, however, we can infer the approximate scale of an object even when an artful photographer omits the ruler-bar of conventional scientific illustration. Certain organic shapes cannot work at very large or very small sizes, because the ratio of surface to volume declines so rapidly as objects grow. Thus, for example, large terrestrial animals must evolve very thick legs,

Skulls of Flathead Indians, skull from China

with strength measured as cross-sectional area, to hold up the body's volume.

Sphericity, however, remains an option from minute ball bearings to Jupiter. Are the objects in this photograph baseballs with stitching, skulls with sutures, or planets with river courses? Clearly, these spheres cannot be baseballs, for the meandering fractality shown here by natural sutures contrasts strongly with the simpler regularity that must be imposed by limits and efficiencies of human stitching machines. But how do we know that the near-perfect sphere, upper right in the photograph, is a human skull, something that we can encompass in our grasp, and not a whole world that only God could hold in his hands?

First, a reason based on later attachments. These skulls come from the anatomical collection of the Harvard Medical School, and museum curators love to affix identifying labels to their objects. Our potential Mars bears two curatorial labels. Some New Age cultists may think that they can see a carved human face on a Martian mountain, but even the most gullible ufologist will not argue for such giant labels—so this object must exist at a human scale.

Second, a reason contemporaneous with the skull itself. Gravity dictates that spinning objects at planetary sizes must be nearly spherical. (Gravity, as a weak force correlated with mass, scarcely affects small objects, but does control massive bodies.) The other two objects depart too far from sphericity to represent planets, but they make sense at human dimensions. In fact, the form of the lower specimen originated as a work of art, for the owner of this skull belonged to one of the Native American groups that shaped the heads of their offspring in infancy by long-continued pressure against swaddling boards.

This flattening can be imposed only because the human skull remains so flexible at birth (leading to the temporary "banana" shape of a just-born head). The main reason for flexibility lies in another feature that helped us to distinguish the proper scale of this photograph—the sutures, which remain pliant and unossified at birth. If the fetal head did not possess this flexibility to fit through the tight squeeze of the birth canal, human mothers could not bring such brainy babies into the world. But sutural mobility cannot represent an adaptation for limited space in mammalian birth, because our reptilian ancestors, who must only break free from a sufficiently roomy egg, also possess the same flexible sutures. Evolution often relies on such lucky breaks, as I did, in this essay, in using the sutures and their consequences to identify proper scale.

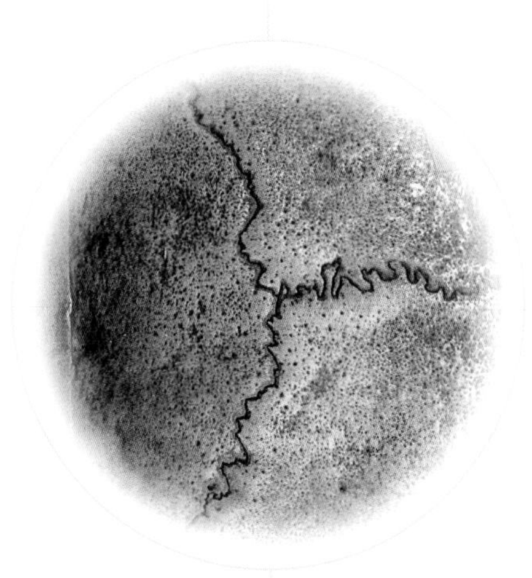

6

big birds

THE JUXTAPOSITION OF THE TWO EGGS IN THE UPPER photograph—one belonging to a songbird of ordinary size, the other to an ostrich, the paradigm of avian largeness—seems to parody the famous epistolary statement of Isaac Newton to Robert Hooke: "If I have seen farther it is by standing on the shoulders of giants." Yet the songbird would pass for the ostrich if I used for my topmost "dwarf" the equally well formed and calcified egg of my favorite land snail, *Cerion*.

This relativity of scale recalls the famous verse of Jonathan Swift:

So, naturalists observe, a flea
Hath smaller fleas that on him prey;
And these have smaller still to bite 'em;
And so proceed ad infinitum.

And I need hardly remind readers that the same Mr. Swift provided our culture's standard example of this intriguing phenomenon by making an

Egg of small bird atop ostrich egg

Egg of Aepyornis *(elephant bird) at left; ostrich egg on right*

unaltered Lemuel Gulliver either a giant or a dwarf—depending on his temporary residence in Lilliput or Brobdingnag.

If you think my example ill chosen because ostriches form the topmost boundary among birds—and therefore represent an absolute standard for earthly immensity—consider the second photograph, showing the same ostrich egg as a runt compared with an egg of *Aepyornis*, the extinct elephant bird of Madagascar. Some species of these flightless ground birds stood more than ten feet tall. The largest intact fossil egg measures about three feet in circumference, and held a volume in excess of two gallons.

To extend this photographic observation into a generality (at least for birds), may I suggest an overall revision to our conventional view of prototypical birds as small and delicate creatures that fly and go *tweet*. Aerial life has emerged as the great success story of avian evolution, and most of the 8,000 to 9,000 species of modern birds are flyers—and therefore constrained to be small. But, for at least five reasons, we should not view large size (and consequent life on the ground) as uncommon or anomalous in birds.

First, birds almost surely evolved from creatures unsurpassed as symbols of bulk: dinosaurs. To be sure, birds arose from a line of agile, running carnivores, the smallest of all dinosaurs—and not from *Tyrannosaurus* or any other monster honored by pop culture. But even the smallest dinosaurs vastly exceeded the average size of a modern flying bird, and our contemporary tweeters therefore evolved by substantial decrease of ancestral size.

Second, large ground birds remain common today, though underappreciated by northerners because all of these species inhabit the Gondwana, or southern-hemisphere, continents—ostriches in Africa, rheas in South America, emus and cassowaries in Australia and nearby islands. Third, the two largest and most spectacular groups of ground birds have only recently become extinct—the elephant birds of Madagascar and the moas of New Zealand, both probably extirpated by the first human settlers of these islands.

Fourth, birds only narrowly missed an evolutionary opportunity to become dominant carnivores on land. Some 45 million years ago, before the advent of large carnivorous mammals, giant ground birds ranked as major predators of herbivorous mammals on northern continents. Fifth, we have no reason to suppose that mammalian carnivores won by predictable superiority rather than quirky chance. On South America, an isolated island continent until the Isthmus of Panama rose just a few million years ago, another group of carnivorous ground birds (the phorusrhacids) persisted and flourished (leaving the large local mammalian carnivores, all marsupials by the way, in second place) until the link with North America forged a route for modern big cats—the jaguars and their relatives—as usurpers.

Big Bird should be judged as a badge of natural honor and substantial currency, not only as an overstuffed chicken bringing welcome enlightenment to the wasteland of children's television.

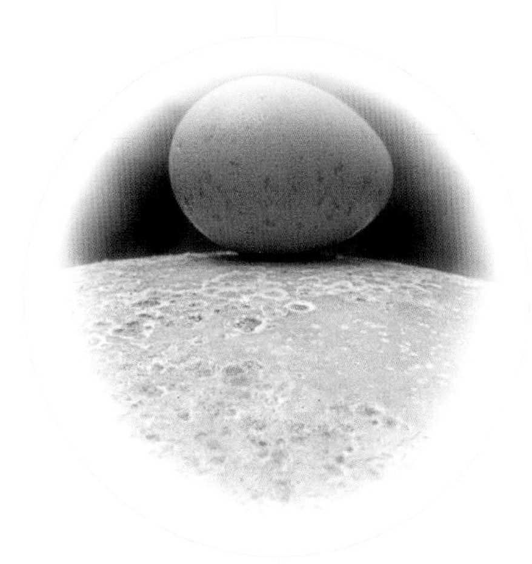

inside, outside—
right, left

7

exotic interiors

DISTANCE ANYTHING FROM THE COLD LIGHT OF EASILY repeated direct sight, and people will construct legends galore about wonders and oddities in strange places. This inventive principle applies especially well to natural history, in endless tales of peculiar creatures from distant and rarely visited lands. Linnaeus, impressed by travelers' reports about exotic humans with hairy bodies and bushy tails, established the subspecies *monstrosus* to place these fanciful wild men among the races of *Homo sapiens*. But to cite an old cliché, truth really can be stranger than imagination, and genuine creatures from maximal distances may be even more exotic than our mentally constrained fictions—whether the distance involved be spatial, as on our modern earth (giant tube worms recently discovered at deep-sea vents), or temporal, as in our fossil record (the peculiar creatures of the Ediacara fauna, the earth's first community of multicellular animals evolved some 600 million years ago).

But distance cannot always be calculated in terms of tape measures. The physically nearest, as Tennyson told us, may lie farthest from our sight or

Curious creatures from Aldrovandi's
History of Serpents and Dragons, *sixteenth century*

influence. No situation in all biology can be more ironic than our inability to see what resides only inches away, but *inside* our bodies (see essay 1). Our skin's opacity has stymied medical science from the very beginning, and our new technology of CT scanners and MRI imagers has produced a revolution (and saved millions of lives, including my own). Thus legends and realities about foreign creatures within human bodies belong in the same category of exotica as the products of physically distant places. Moreover, since we maintain a special sense of horror about animals *within* us (how else to explain the success of such a grade-B fright film as *Aliens*), tales about internal parasites elicit a heightened frisson of wonder and disgust.

These five woodblock figures appear together on a single page of *The History of Serpents and Dragons*, a late-sixteenth-century work by the greatest chronicler of natural history in the Renaissance, Ulisse Aldrovandi of Bologna (1522–1605). All these drawings represent creatures that supposedly dwelled within a human body but exited to our sight by a variety of unplanned means: expelled from a uterus for number 1, in vomit for number 2, in urine for number 4, and out the other major excretory orifice for number 5. Number 3, labeled a "monstrum anguineum," or eel-like beast, must represent the devil's perverse counterpart of a truly blessed event—for this creature, Aldrovandi writes, exited "ab alvo virginali," from a virgin's womb.

Are the true and troubling parasites of *Homo sapiens* any less strange than these fictions? Consider the twisting, almost textual

Wax model of tapeworm found in human intestine

flow of the twenty-foot wax tapeworm shown in this photograph. The specimen, also from Bologna but a replica of reality this time, dates at least to the early nineteenth century. A list compiled by papal command in 1825 to assess Napoleon's despoliation of cultural treasures in lands then ruled by the Holy See includes this specimen (Napoleon surely showed good taste in deciding what to plunder and what to leave behind).

Tapeworms are simple but devilishly effective creatures, members of the most "primitive" phylum of "higher" animals (Platyhelminthes, or flatworms) on those falsely construed "ladders of life" that we all memorized in high school biology. (Tapeworms grow neither mouth nor digestive tract and absorb food through the external body surface.) Humans ingest the tapeworm at a small, early stage, when the parasite resides in the muscle of another host, usually a cow or a pig. The scolex, or head, then attaches by suckers or hooks to the inside of a human intestine, where it begins to bud off a long chain of segments, called proglottids (up to twenty or thirty feet worth in some cases, thus snugly filling the entire length of the intestine). The proglottids are reproductive factories; each may house up to 40,000 embryos. Tapeworms, as hermaphrodites, can fertilize themselves. Proglottids may break off and exit the body in feces, ready to be ingested and to start the grisly cycle again. Unpleasant (for us) to be sure, but God works in many ways . . . and we sometimes meet our Waterloo from within.

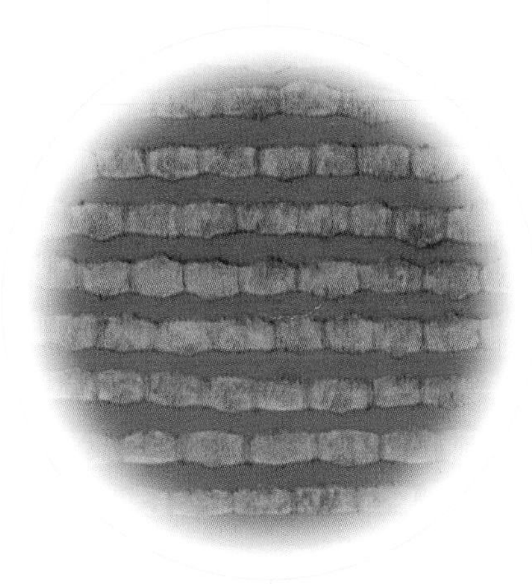

Tin Boy

8

the allure of equal halves

IN COMPLEX AND MOBILE ANIMALS, BILATERAL SYMMETRY—the division of the body into two halves, each the mirror image of the other—dominates all other patterns of organic architecture. Biologists tend to explain such preferences by arguing for the adaptive value, or Darwinian utility, of favored features, and I do not doubt that construction in two equal parts offers great advantages to moving creatures, who can then keep their sensory organs up front, trail their waste products behind, and face the dangers and opportunities on both sides without bias. (By sensible contrast, most immobile organisms, from flowers to coral polyps, build their bodies in the radial symmetry of a circle, where any diameter will divide the creature into two equal halves. No spot on the circle can claim any preferred status, and the organism can feed and respond to danger from all directions.)

I accept these venerable arguments, but we must also acknowledge a different kind of architectural reason—well illustrated in this photograph of an old toy, a tin boy that spent many years eroding in a junkyard. Bilateral symmetry also permits an easier construction of complexity, enabling two halves to be built for the price (so to speak) of one—that is, by a single rule endowed with a switch for mirror reversal. (The tin boy illustrates such a process, with the midline between his two halves marking the junction between the two parts of his mold.)

The subsequent history of tin boy's erosion records a second key property of bilateral symmetry, or at least of human perception. We are so accustomed to noting bilateral symmetry in complex animals that we experience an eerie feeling when creatures depart from this norm. Thus a South American monkey, using its large tail as nimbly as its four limbs and moving asymmetrically "on all fives," tends to disorient us. And the sideways motion of crabs has been noted as anomalous ever since Aesop.

This preference for symmetry may be deeply encoded into our evolved emotions, as many studies show intrinsic preferences (in humans as well as in many other animals) for the most symmetrical mating partners—a sensible Darwinian strategy for choosing "better halves" more likely to lack serious genetic blemishes. The tin boy's two halves have eroded differently during his years of exposure—and this disparity strikes us as odd, if not sinister.

And yet, complex nature remains chock-full of exceptions to bilaterality, and many organic phenomena show distinct "handedness," or strong preference for one form to its looking-glass counterpart. Indeed, life itself displays such asymmetry: organic amino acids, the building blocks of living matter, all adopt the "left-handed" configuration, whereas inorganically produced amino acids arise with equal numbers of left- and right-handed molecules.

This asymmetry also rules the human body in two major ways. First, our exteriors may be tolerably bilateral, but many interior organs (especially the heart) lie to one side of the body's midline, and also depart from bilateral form in their own construction. In the best-known case, the two halves of the brain differ only slightly in external anatomy, but become markedly disparate in neurological function. As a validation for my earlier claim that bilaterality may be built by simple rules, several genetic mutations can provoke full reversals of the usual order—snails coiling left rather than the usual right (for most species), or human internal organs forming on the "wrong" side, a condition known as *situs inversus*, or "reversed position."

Second, we impose behavioral asymmetries upon bilateral bodies. Right- and left-handedness represents our premier example. No one knows why right-handers predominate, but our lamentable disparagement of the unusual stands forth in the etymologies of all common European languages—as righties are dexterous (from the Latin *dexter*, meaning "right"), while our poor tin boy looks sinister in his asymmetry (from the Latin *sinister*, meaning "left").

9

pulling teeth

WE DEFINE OUR OWN CLASS OF VERTEBRATES BY THE growth of an internal skeleton (while insects and many other invertebrate groups form their protective hard parts on their outsides). And we tend to be fascinated by what we cannot usually see, yet recognize as essential. Thus we take delight in X rays of our own invisible interior, experience a special horror if bone accidentally breaks through skin, and symbolize the macabre as a dance of articulated skeletons. Yet we rarely realize that one important part of our skeleton, representing the oldest structures of all by evolution, stands readily exposed to daily scrutiny and attention—our teeth. Ya gotta eat.

But necessary exposure implies inevitable risk. Our bones may be cushioned, but our teeth remain vulnerable to all the slings and arrows (and fists, sidewalks, and hockey pucks) of the world's outrageous fortune. Modern dentists possess an armamentarium of sophisticated devices for preservation and salvation. But their forebears could do little more than follow, in a literal manner, the prescription that Jesus dictated (metaphorically, I trust) as treatment for a wandering eye (in a man with an adulterous heart) in his Sermon

The Dentists: Meissen porcelain figurines, eighteenth century, against nineteenth-century French print of the dentist as torturer.

on the Mount: "if [it] offend thee, pluck it out, and cast it from thee, for it is profitable for thee that one of thy members should perish, and not . . . thy whole body."

As one of the few routinely successful forms of old-time surgery, and as a crude and unpleasant way to save many lives that would otherwise have been lost to spreading infection, the yanking of teeth became an obvious butt for humor—as shown here both in the elegance of eighteenth-century Meissen porcelain and in a background print of the same age and activity. (I don't imagine that the Meissen folks crafted many scenes of gangrenous death on the battlefield or in the birthing room, where surgery could do little and medicine usually made matters worse.) Nonetheless, as depicted in both print and porcelain, the subjects were not amused; "like pulling teeth" remains our standard metaphor for necessary things done with great reluctance.

Nonhuman mammals often face a precisely opposite dental problem—an upper limit set to life because teeth can neither be removed nor replaced. Non-mammalian vertebrates (including sharks as a primary example) can replace worn or broken teeth. But we mammals pay a price for each supposed gain in adaptation. Mammals

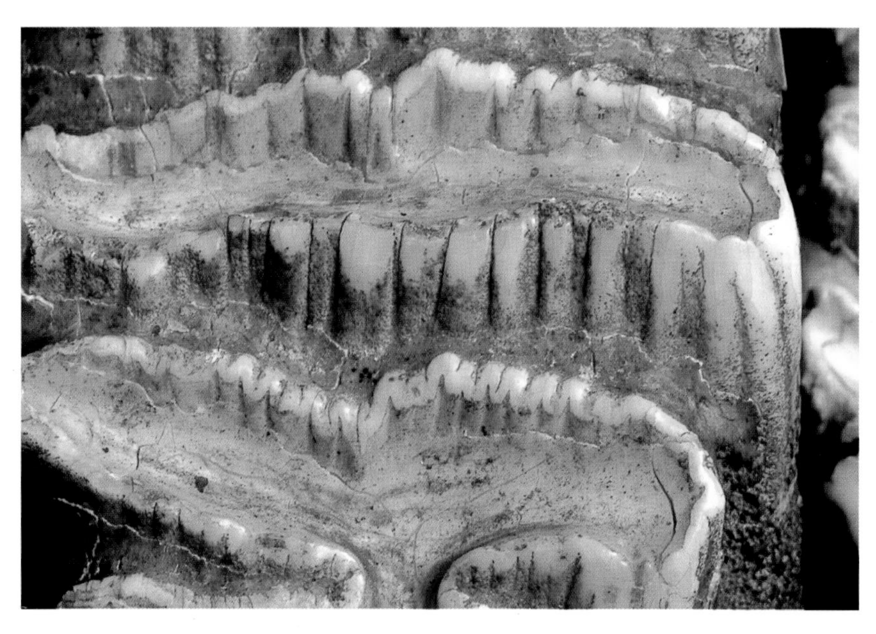

Molar of mammoth

have evolved complex teeth, requiring delicate and precise occlusion between upper and lower members in chewing. Good occlusion implies permanent positioning, so mammals must give up the conveyor-belt system of continuous replacement used by fishes and reptiles with simple pointed teeth, and settle instead for only two sets—"baby teeth" for the small juvenile jaw and a single adult set for later life.

In Darwinian nature, life may generally follow Hobbes's dictum of nasty, brutish, and short, but some mammals survive to die of old age. Ironically, the most common cause of such "programmed" death seems to be the wearing down of irreplaceable teeth to decrepitation. Ya gotta eat.

Adult elephants have no potent nonhuman predators. Their large and elegantly complex molar teeth—shown in surprising scale in the photograph—do not follow the usual mammalian rule. Only one molar can fit on each side of the jaw at a time—for a full complement of four molars at any moment. A worn-down molar can be discarded and another emplaced (whereas most adult mammals sport all their molars, usually two or three for each side of each jaw, as permanent fixtures). But after the last elephant molar gets emplaced and worn down, no substitutes remain, and a toothless old animal may starve to death.

How ironic, then, that modern elephants suffer no threat as a species by dint of this natural form of dental distress, for all elderly animals must eventually die. Rather, elephants have become vulnerable because we have imposed upon them, for rapacity rather than relief of distress, the uniquely human strategy of pulling teeth. We kill them to extract their incisors for ivory.

Restoration of Pygmalion

10

layering

MANY LAYERS OF COMPLEXITY COVER BOTH THE INTEGRITY of ordinary people and the dissembling of cowards. In the most physical example, our relatively smooth skin hides a wondrously ordered intricacy within, from vessels and organs down to the "bare bones" of the framework. We then add cultural overlays, both physical and psychological, upon this biological barrier—by clothing our physical nakedness (an image with deep roots in Genesis), and by masking our raw emotions with conventions of civility and decorum.

We cannot live (and society cannot function) without these covers, yet we also bemoan the sequestering of a supposedly true self. Our metaphors for "stripping away" either celebrate a raw reality or castigate an exposed corruption. "Naked came I out of my mother's womb, and naked shall I return thither," says Job. But the belly of the beast also lies under the skin of the saint.

I do feel sorry for Daniele da Volterra (1509–1566), a fine Mannerist painter and disciple of Michelangelo, but known almost exclusively as

Il Braghetone ("the breeches maker") because he followed Papal orders and painted pants and draperies over exposed genitalia in Michelangelo's *Last Judgment* in the Sistine Chapel. But if *Il Braghetone* earns our ridicule for violating Job's precepts about truth in primal advertising, a shroud over our nakedness can win plaudits in other cases—when, for example, we wish to impose uniformity atop the uniqueness of each person (as in a military parade, a symphony orchestra, or the Rockettes on line at Radio City), or when I need more than my love to keep me warm.

This contrast permeates several levels of the two photographs here—an amusing cover of cultural prudery in one case versus a welcome biological shield in the other. (I am also struck by the paradox of using photography to record what photography cannot capture. We must often draw and model the human body because a photograph of a living person cannot reveal the internal details that both art and science need to portray. We can then photograph these models to complete the double whammy of visualization plus dissemination.)

Roger de Bussy-Rabutin (1618–1693) may deserve his primary reputation as a military adventurer and libertine, but his literary abilities struck his countrymen as sufficiently refined to merit his election among the "forty immortals" of the Académie Française. (His most famous work, *Histoire amoureuse des Gaulles*, including four tales of courtly life and loves, led to his imprisonment in the Bastille for more than a year when his enemies charged him with disrespect toward King Louis XIV.)

The photograph of a painting from Bussy-Rabutin's château in Burgundy depicts Pygmalion, the sculptor who, in Ovid's *Metamorphoses*, falls in love with his own statue of a woman. The painting may record both Bussy-Rabutin's status as a former libertine and his quest for respectability in older age, after his release from prison. The Pygmalion of the top layer seems chaste enough, despite his evident ardor, but part of the painting has been stripped away to reveal an earlier, more salacious version, with the sculptor's left hand resting on his statue's breast. In the comic result made of juxtaposed layers, Pygmalion sports three arms. Under the red tunic of revised respectability, we can still trace his old pathway toward copping a feel.

The eighteenth-century Spanish wax model of a woman, from the Museo Anatómico of the Medical School in Valladolid, may disturb modern sensibilities, but represents a common teaching tool of a grittier age that knew death and corruption as part of life's daily passage. Still, I cannot help thinking that, even in 1800, medical students must have shuddered at the graphic realism of internal parts not meant for human sight—and must also have felt a pang of pain and compassion for the beauty of a young woman's unblemished skin covering exactly half a face.

Never doubt the power of art to trump both the limits of empirical reality and the strictures of human censorship. And score a final point for Michelangelo, who painted the apostle Bartholomew right in the center of his *Last Judgment*. Following the convention that martyred saints carry their mutilated parts, Bartholomew, who was flayed alive, holds his entire skin in his left hand. But unlike the Spanish wax woman, who must reveal her internal anatomy under her flayed half, Bartholomew, ready for resurrection, rises in a fully reconstituted skin, disturbed only by Volterra's *braghe* subsequently draped over his privates. In a lovely final touch, Michelangelo painted his own face upon the saint's flayed and former covering—perhaps in hopes of his own forthcoming resurrection, but more likely in mockery of the censorious flaying that all artists of integrity must suffer if they wish to remain whole.

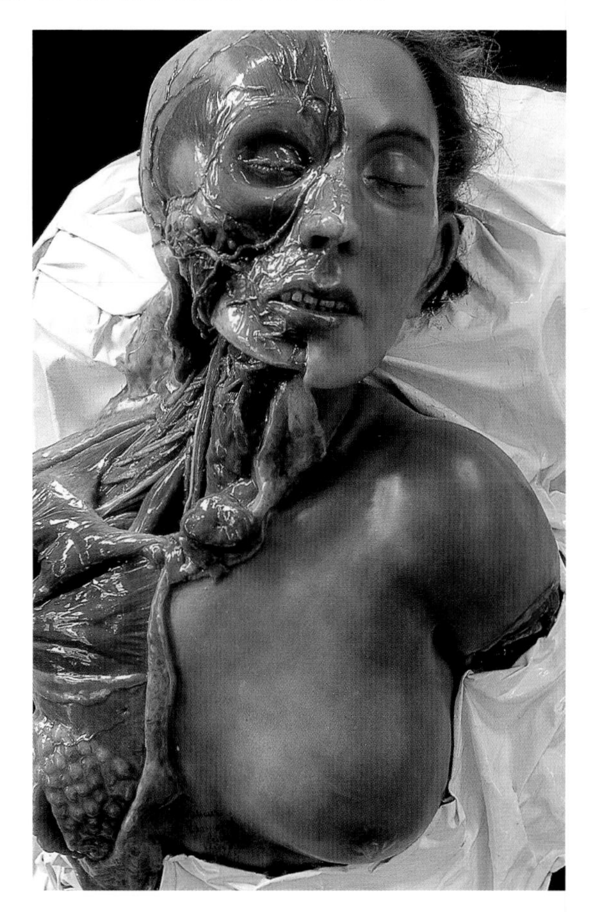

Dissection in wax

richness and loss of information

Polished side of a slab of ammonites

Unpolished side of the slab of ammonites

11

two sides to every issue

WE DELIGHT, AS A SPECIES, IN BAUBLES, BANGLES, AND beads—rich patterns to ornament a surface. Consequently, stones chockablock with fossils have long been favored as polished slabs and columns for buildings and furniture. Purbeck marble, used to build the fluted columns of several medieval English cathedrals, contains little more than masses of fossil snails, packed solid. The black paving stones found throughout southern Sweden feature long, graceful nautiloids; a beautiful specimen points to Linnaeus's name on his tombstone in the cathedral of Uppsala. Have we not all, during pensive moments with a purpose, admired the fossils in marble walls of bathrooms in our better department stores and restaurants?

These two photographs depict the before-and-after in our standard procedure for smoothing slices of fossil-bearing stone. We see both the polished and unpolished sides of a single slab chock-full of small ammonites from the lowermost Jurassic period (the middle years of the reign of dinosaurs on land). In reducing the dimensionality of this specimen from the three of nature's rock to the two of human utility as a polished slab, we both gain and lose information about the fossils contained within.

We gain because the plane of cutting intersects many specimens and reveals interior details not visible on the rough side. (The fossils are harder than the surrounding muddy matrix. The rough side represents a quarrying of the rock from its original setting. The rock tends to break at the junctions between shell and matrix, thus producing a wondrously complex surface, but revealing no interior detail of any fossil.) From the exterior alone (on the rough surface), we note that the shells coil tightly in a single plane. The specimens are obviously mollusks, but they could be either snails or ammonites (coiled cephalopods related to modern squids, octopuses, and more closely, the chambered nautilus). The polished side cuts through several specimens and shows the interior chambers, thus identifying the specimens as ammonites.

Coiled cephalopods live in one chamber at a time. As the shell grows, the animal moves forward and then builds a new wall at the front edge of the chamber most recently vacated. At the animal's death the shell fills with gas and floats upon the sea. In his celebrated poem *The Chambered Nautilus*, Oliver Wendell Holmes used both these features of cephalopod biology to make his metaphorical statement about spiritual progress through human life. First, growth by successive chambers (in a line later borrowed by Eugene O'Neill as a title for a great play):

> *Build thee more stately mansions, O my soul,*
> *As the swift seasons roll!*

And, finally, the soul's liberation after sloughing off its mortal coil:

> *Till thou at length art free,*
> *Leaving thine outgrown shell by life's unresting sea!*

But we also lose information in the two-dimensional polished version, primarily because most of us cannot properly visualize a three-dimensional object from the limited evidence of a random cross section cut at an odd angle. When an ammonite shell lies on its side, we have little trouble with a two-dimensional version, for this orientation matches our traditional drawings. But when the shell lies on its edge, or at an oblique angle to the plane of the polished slab, then the cross section resembles a line of rings (the ridges on the shell's exterior) or a blob shaped between a crescent and a hemisphere (for oblique cuts at various angles). The shells can be identified more easily on the rough side because they either stick out above the surface or leave a well-formed impression in the matrix.

We can learn more about the life of these ammonites through the circumstances of their death (a branch of paleontology called taphonomy, or the study of burial). If the shells had been deposited in a gentle current, nearly all would have fallen in their most stable position, that is, on their sides. But the shells in these slabs lie every which way, indicating a more tumultuous site of death. And why are these rocks made entirely of ammonites? Where are all the other creatures that must have shared their sea? Were the currents that transported these shells to their burial site so uniform that they carried objects only of particular weights and shapes—and were these ammonites the only available creatures with such properties?

Any block of rock encodes complex messages. Shakespeare, after telling us that we might make sweet use of adversity, then stated, in the next lines of *As You Like It,* that our life

> *Finds tongues in trees, books in the running brooks,*
> *Sermons in stones, and good in every thing.*

12

report to the librarian

BACK IN THE GOOD OLD DAYS OF LESS EXPENSE, I ONCE encountered a sign on the beverage machine in my office: "You can't get your dime back by pouring the coffee into the machine." The second law of thermodynamics sets time's arrow—and this universal direction leads to accumulating disorder measured as increasing entropy. For people engaged in the sciences of history (I am a paleontologist by trade) the natural deterioration of former order acts as a persistent enemy and a perennial challenge.

This photographed construction of the surprising beauty in an intellectual's nightmare shows nothing but books—or, rather, the fragments and transformed exuviae of books after long exposure to the elements in a junkyard. We see spines, cover cloth and cardboard, and, in two clumps, the fused and wadded remains of pages, now matted together like lumps of clay.

Amid all this deterioration, one label stands out in an almost pristine state, mocking our hopes for preserved and eternal order: "Keep This Book

Keep This Book Clean (construction)

Clean. Do Not Turn Down the Leaves. Borrowers finding this book pencil-marked, written upon, mutilated, or unwarrantably defaced, are expected to report it to the librarian." Well, Mrs. Hochsburg (the stereotypically stern librarian at the storefront branch around the corner during my childhood), I want to report nature for all these sins and more—and just what are you going to do about it?

Among the lugubrious scenarios, apocalyptic and otherwise, that visionaries and intellectuals have always constructed for the end of the world, a new genre appeared and gained prominence after the second law of thermodynamics pierced popular culture in the late nineteenth century: the heat-death of the solar system, an exhausted sun burned out and all the planets entropically equalized and icebound. But I suggest that we look at the pristine label on the coagulated book and rejoice for two reasons.

First, the historian's salvation. In general, evidence degrades and disappears. The vast majority of organisms leave no trace of their existence whatever, and our fossil record preserves the pieces and activities of a tiny minority. But sometimes, against all the power of time's decay, a precious item from the past resists destruction and stays put in all its detailed glory. We learn so much from these perfect fragments, and their pristine preservation amid so much deterioration seems to mock entropy with panache, thus thrilling our senses and lightening our spirits. We rejoice at the Pantheon's unique Roman roof; the Book of Isaiah on a Dead Sea Scroll; the feathers of *Archaeopteryx,* the first bird, preserved in rocks more than 200 million years old. These improbable survivals grant us hope that we can decipher the distant past, even down to the lovely details that make history—just as the perfect preservation of a librarian's orderly label, on a book otherwise coagulated to mush, stands defiantly against entropy.

Second, our local hope. Entropy must increase, and order deteriorate, within closed systems infused by no outside source of energy. But the earth, throughout 4.5 billion years of history, has not existed as a closed system and can therefore act as a temporary stage for local and systematic increases of complexity. Our planetary home remains open, because, at every moment, the sun pumps in massive amounts of energy that can be converted into increasing order. (Yes, the entire solar system may be effectively closed, and the sun will eventually burn out and die—but why lament such incredibly distant inevitabilities? Incidentally, the next time your local creationist tries the notoriously phony argument that evolution cannot be true because natural order must decrease, tell him about open and closed systems.) The library of this particular label may be extinct, but the energy of all the world's librarians can keep this grandest of intellectual institutions vigorously alive, fully expanding and eminently in order, until the sun burns out!

13

the old red sandstone

WE SCIENTISTS ARE NOT SO HEARTLESS A LOT THAT WE always designate nature's phenomena either by numbers or with mysterious Latin names longer than the mite or muscle thus identified. We are not immune to occasional whimsy; numbers may come in googolplexes, and quarks with charm help to build the universe.

The names of the geological time scale—Cambrian, Ordovician, Silurian . . . up to Holocene—do not, unfortunately, lie among the felicitous entries. But we make up for an uninspired framework by choosing some memorable monikers for certain rock formations within these periods of time. The British sequence deserves particular praise for preserving, even as technical terms, some wonderful old names taken from the vernacular of farmers and quarry-men. The soils of the Cornbrash, an official division of the Jurassic, can grow your wheat. The Millstone Grit then provides excellent stones to grind the resulting grain. But I don't know what one does with the Clunch Clay.

But the prize for the best name must be awarded to the Old Red Sandstone, a quarryman's term for Devonian deposits, often coarse and reddish in color, that underlie the economically crucial Coal Measures. Woe betide any quarryman who can't tell the difference between the Old Red and the admittedly similar rocks of the New Red Sandstone, which *overlie* the deposits of coal. (If you misidentify Old as New and drill down hoping to find coal beneath, you may dig to Satan himself and never extract any usable warmth.)

The Old Red also houses some exquisite fossil fishes, representing some of the earth's earliest vertebrates. (Since these fishes had become extinct by New Red times, the fossils can also be used to make the crucial distinction between these two formations.) Hugh Miller, Scottish stonemason turned geologist, and one of the most popular of mid-Victorian writers, called his first book *The Old Red Sandstone* (1841). As a paean to the self-help movement, Miller urged workingmen to vent their frustration against the rich by improving their own minds—for example, by studying the fishes of the Old Red Sandstone. Miller writes:

> You are jealous of the upper classes; and perhaps it is too true that, with some good, you have received much evil at their hands. . . . But upper and lower classes there must be so long as the world lasts; and there is only one way in which your jealousy of them can be well directed: do not let them get ahead of you in intelligence.

Louis Agassiz, who later came to America, where he founded the Museum of Comparative Zoology at Harvard University, established his career by writing a beautiful and comprehensive monograph on all the world's fossil fishes—*Recherches sur les poissons fossiles* (1833–1843). The five volumes of wonderful plates represent some of the first examples of color, or chromolithography.

This fish, *Cephalaspis lyelli,* named for the great geologist who befriended Agassiz during his visits to England, belongs to the class Agnatha, the jawless ancestors of all later vertebrates. Rosamond Purcell located the original fossil at the British Museum, and we

Fossil fish (above) and lithograph of the same from Louis Agassiz's Poissons Fossiles

present a comparison of Agassiz's lithograph (drawing on stone) with the actual specimen in stone.

This comparison shows the advantages of Agassiz's new technique in scientific illustration. The subtly variegated colors of the Old Red had previously been intractable to anything but laborious hand coloring. But by overlaying colors of slightly different hues one screen at a time, Agassiz could capture the changing textures of the Old Red. However, the differences between illustration and prototype also catch our attention. The specimen lies *in* the rock, but the drawing appears *on* the rock. The drawing lacks depth but provides greater clarity of outline. To this day, most naturalists prefer drawings to photographs for presentation of anatomical complexity and detail—and good scientific artists remain in high demand.

Agassiz lavished great care on his illustrations, both for aesthetic pride and for solvency, since he sold the volumes by subscription, largely to lords and ladies who loved the pictures and cared little for the complex text. But Agassiz eventually became frustrated by the unreliable work of subcontractors, and he therefore established his own lithographic press. The bankruptcy of this enterprise, and the offer of a fat lecture fee from the Lowell Institute, inspired his migration to America, thus launching one of Boston's proudest intellectual lineages.

superposition

WE MURDER TO DISSECT," WROTE WORDSWORTH, BUT I find his defense of totality altogether too encompassing. I would never denigrate the virtues of wholeness, but who can deny both the scholarly and aesthetic pleasures of decomposing integrity into components, and then identifying the order behind their amalgamation?

Principles of order vary according to discipline and object, but geologists (and historians in general) favor a study of "overlay" to determine temporal order as the chief tool for unraveling a complex ensemble. Simply put, thing one must be older than thing two if thing two overlies or otherwise modifies thing one in a manner implying the initial presence of thing one alone. Geologists tend to be quite literal about "overlying." We determine temporal order by the so-called law of superposition—strata on top of a pile win the prize for youthfulness because the bottom layers must have been deposited first.

This fossil fish from the Solnhofen limestone of Germany illustrates the power of overlying as a criterion for understanding the genesis of a complex

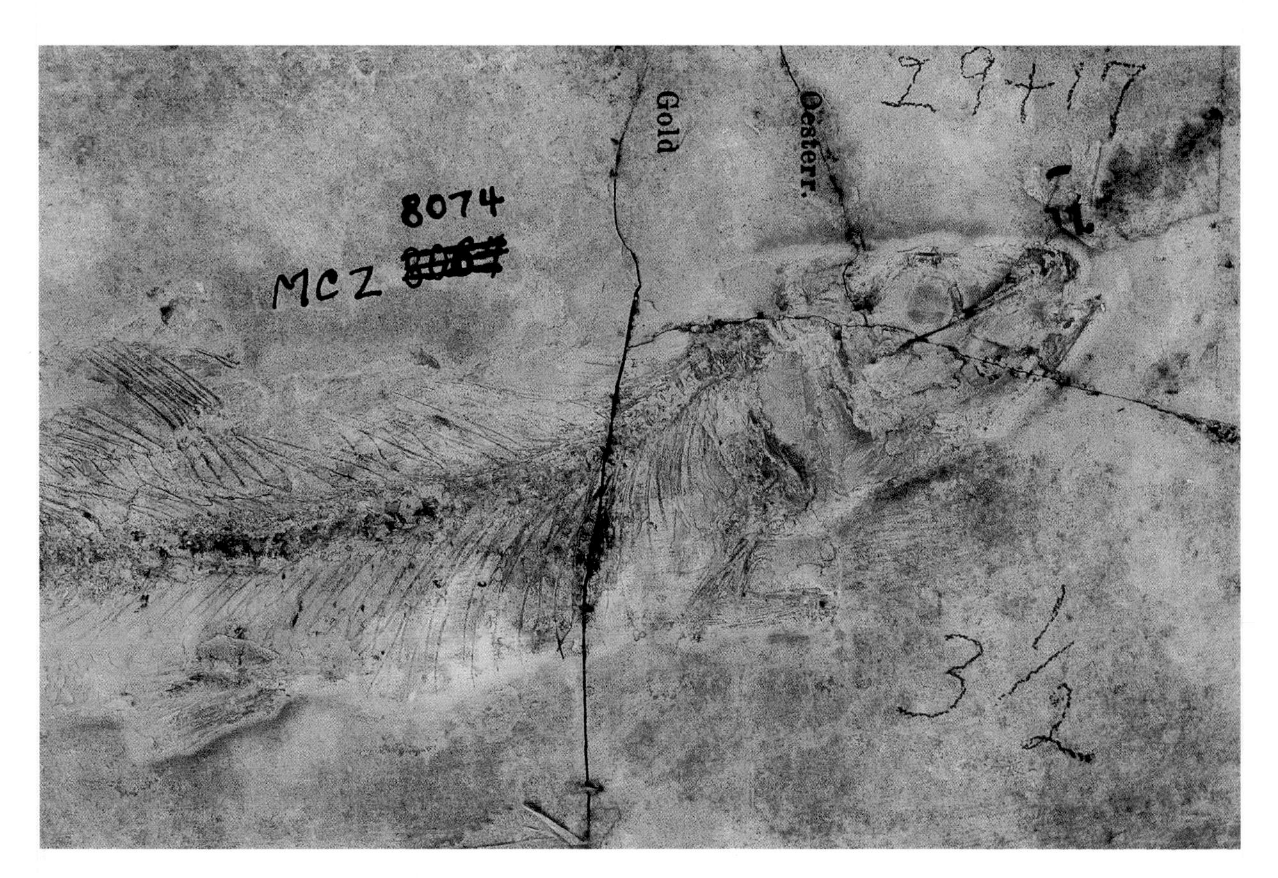

A well-examined fossil fish from the Solnhofen limestone

totality. We note a sequence of layers, from inside to outside, and from deep embedding to barely adhering superficiality. We see, in the fish itself, both inside (the bony skeleton) and outside (the form of the fins and body outline, since soft parts can be preserved in the exquisite Solnhofen fauna). The cracks overlie both fish and surrounding rock and must therefore postdate the fossil's formation. Similarly, and obviously in this case (for we understand human interventions better than nature's ordinary ways), the metal staple suturing the main crack below the fossil must represent the youngest stratum of this sequence. Atop this natural ensemble of rock, fossil, and cracks, we note a variety of human overlays in two forms: writing directly upon the stone, and adhering paper. Penciled numbers to the right (29 + 17 and 3½) may mark the oldest additions, and may represent measurements or inventories. The inked number is a Harvard Museum identification (MCZ for Museum of Comparative Zoology) —and here we note overlay within overlay, for the first number has been crossed out and replaced by a second, obviously later, designation.

Finally, consider the puzzling words *Gold* and *Oesterr*, seemingly printed *on* the rock. A resolution for this latest layer required some real detective work. In the museum's drawers of Solnhofen fossil fishes we found several thin nineteenth-century printed lists from the "Süddeutsches Correspondenz-Bureau in München." These lists provide daily rates of exchange for European currency. The categories include gold and currency from Austria (Oesterreich). These sheets must have been used as packing material to separate and protect slabs of fossils when a large collection of Solnhofen material traveled from Germany to America. Parts of the thin paper, pressed by the weight of fossils above, stuck to the stone. Fibers of the paper can be seen clearly under a microscope. (Note also the larger scrap of paper near the fossil's head.)

Intellectual life should not be construed as two cultures of science and humanities at war, or even at variance. Human culture arose from the material substrate of a complex brain; and science and art meld in continuity. The sequence of superposition on this rock—an unbroken transition from things of nature to things of art, flesh to rock to paper to ink—illustrates the embedding of mind in nature.

II

mind and nature

mental biases

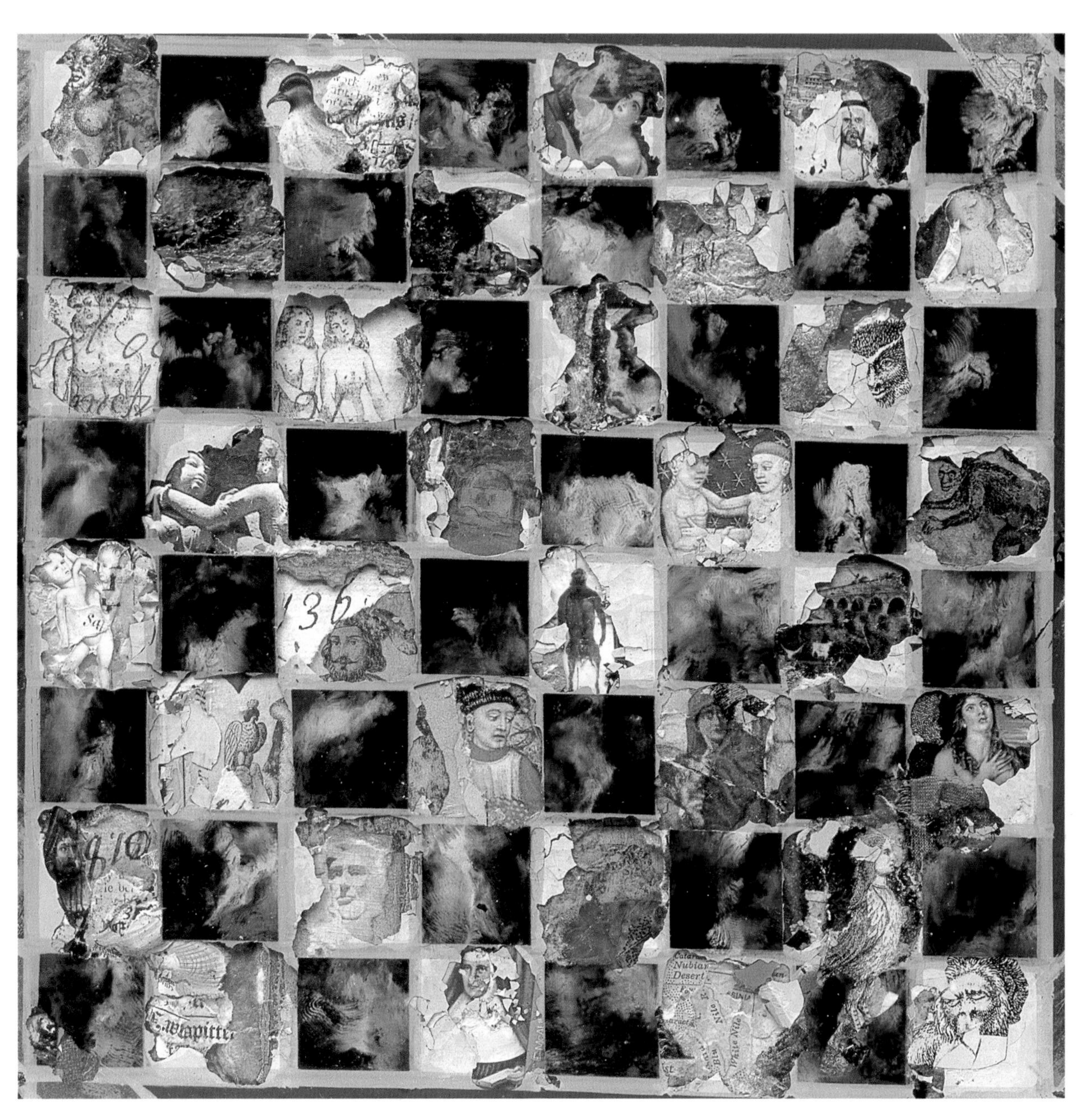

It's How You Play the Game (construction)

15

faces are special

We cannot perceive the world in a totally "objective" manner, for we must filter reality through meshes of physiological preference and social prejudice. I once tried a naïve personal experiment, designed to discover the *Ding an sich* beneath this overlay. I reasoned that, although overt vision must alter and interpret any tableau, the afterimage "seen" through closed eyes might bypass conscious control and record a scene photographically. I stared for a long time at a mosaic bathroom floor with a regular arrangement of black and white tiles—but with a prominent crack through a small section of the pattern. Try as I might, I could not incorporate that crack into the afterimage—as though my retina or brain could only record the regularity and refused to encode the one prominent departure.

Among such biases of perception, none surpasses our genetically encoded tendency to recognize faces. Neurologists have developed good evidence, including numerous studies of cerebral mapping and neuronal firing, for such a useful preference in many mammalian species. This strong perceptual bias both provides benefits and causes troubles. I doubt that we would see faces in the illusionist paintings of Arcimboldo (human portraits constructed from composite fruits, vegetables, fish, or fowl) if we were not programmed to pick out a pattern of midline elements (nose and mouth) and bilaterally symmetrical features (eyes and ears)—and then infer a face from these few clues. But we would also not be tempted to see a face in a Martian monolith (and invent a ridiculous scenario to fuel the headlines of shopping mall tabloids) just because a large oval rock happens to feature a few fortuitous holes in the right places (see essay 5 for further comments).

Purcell has played upon our affinity for faces in her brilliant refurbishing of an old and tattered reverse-painted glass chessboard. She left the dark squares as she found them, but festooned the white squares with a variety of complex overlays, mostly faces, drawn from a wide range of sources: from Buffon's mulatto midget in the initial square (first white square of row one), to Tulp's famous orangutan in the last spot (fourth white square of row eight). In between, you will find (among others) some of history's famous conjoined twins (row two, square eight, and rows three and five, first squares); figures from tarot cards (row five, square three, continued into row six, square two); twins from a medieval book of hours (row four, square six); a nun from an Italian stamp (row eight, square four); a harpie (row seven, square seven); some birds; a few primates; and Honest Abe on Mount Rushmore (row seven, square three).

The more you look, the more you see, as our programmed preferences search out their intended targets. Look hard enough, and you will also find unintended faces in the unaltered dark squares—for oval shapes with appropriate blobs and dots often appear in random arrays, and our mental machinery for recognizing faces then performs the translation (try row two, squares one and seven; or row seven, square four). Looking again, I can find a face on almost every dark square (look at the sultan in row six, square three). Doesn't every cloud bank include a camel?

Yes, faces are special. Even the words find linkage in synonymy—for *face* derives from *facies*, and *special* from *species*, and both Latin words mean, roughly, overt or external appearance. Hence, we derive technical terms in my two fields of geology and biology: *facies* for the local environment recorded by characteristic features of strata, and *species* for the fundamental unit of our taxonomic hierarchy (based on unique and distinctive features, rather than shared properties of larger groups like genera and families). Hence also, the two vernacular words of my title: *face* for a person's most evident and revealing feature, and *special* for a distinctive and particular appearance.

T. H. Huxley described our struggle to comprehend nature as a battle of wits: "The chessboard is the world, the pieces are the phenomena of the universe, the rules of the game are what we call the laws of Nature. The player on the other side is hidden from us." In our ignorance, and with our limitations, we do indeed see nature as through a glass darkly—but perhaps, someday, face to face.

16

placid in plaster

IN 1906 THEODORE ROOSEVELT WON THE NOBEL PEACE Prize for his efforts in mediating an end to the Russo-Japanese War. Also in 1906, the two Chicago baseball teams, the White Sox and the Cubs, won their league titles—and so, for the first time, the so-called "World" Series never left the confines of a single city. This mixture of maximal internationalism and extreme parochiality suffused another, absolutely unknown event of the same year: the manufacture of sixty-four face masks of native people on the island of Nias (near Sumatra in Indonesia, then the Dutch East Indies) by the Dutch colonial anthropologist J. P. Kleiweg de Zwaan.

On the first theme of expansive, romantic exoticism for an international science of anthropology, I can only commend Kleiweg's dedication and industry under difficult conditions. His thick volume of *Anthropologische Untersuchungen* (Anthropological studies), published in 1914, presents more data on measurements of more parts of the body in more charts than I have ever seen before.

But a truly narrow parochialism also pervades Kleiweg's heavy German

Plaster casts of the inhabitants of Nias

and weightier charts. He invested such overweening faith in the power and utility of numbers. What did he think his elaborate calculations of finger bendability and eye-blink frequency could teach us about human diversity? (The Nias people represent an interesting amalgam of Malay, Polynesian, and Melanesian stocks.) Kleiweg's efforts might have reaped rewards if he had known the statistical techniques that can make sense and pattern from such a mountain of data (but the procedures hadn't yet been invented)—or if he had possessed sufficient mental flexibility to alter, by any conceivable interpretation of the data, his firm and canonically European belief in the inferiority of conquered peoples in colonial territories. But I sense no such option in his pervasively biased words.

The face masks in the photograph bear witness to Kleiweg's industry, a common feature of his "just the facts," "more measures the merrier" age of ethnography. Kleiweg describes with zeal how he poured plaster of paris over the faces of his living subjects, and how he took special care, lest material drip either behind the ears (making removal of the mask difficult) or into the open nostrils (a rather more distressing possibility, at least for his subjects). Kleiweg took great pride in his technique. Most colleagues, he states, kept the nostrils of subjects open with goose quills while making masks, whereas he used greater care and could therefore avoid this tickling expedient. As a chief sign of his own confidence, Kleiweg states that the visages of sixty (out of sixty-four) subjects "appear to have a fully peaceful and not at all anxious expression." (Readers may judge for themselves. I detect no predominant complacency on these plaster faces—perhaps resignation or forbearance might provide a more fitting description.)

All Kleiweg's exquisite care in measurement, all his experience among the Nias for several years, never led him to question any of the conventional presuppositions, then held by nearly all Western scientists, about the ranking of races. At best, Kleiweg depicts the Nias in a paternalistic mode, as "a carefree, lively, happy, and childish people." At worst, he excoriates their "carelessness, which has, for consequences, laziness and uncleanliness." He seems especially offended by their unreliability ("I have never known a people who lie so much") and their failure to understand time and keep appointments ("a characteristic which makes work with them extraordinarily difficult").

These masks, the major material result of Kleiweg's years of labor, represent his reward—and mark his failure. Kleiweg finally overcame the "unreliability" of his informants. He caught them all in plaster, and they could never keep him waiting again. With their placid visages they look out at him forever.

But Kleiweg's work has been totally ignored and superseded—intellectually and morally. Purcell found his masks in storage in museum boxes (where they will and should survive, but mainly as illustrations of the history of anthropology). I found his printed work in the stacks of Widener library at Harvard. His book had stood on the shelf since publication in 1914. It had never before been checked out.

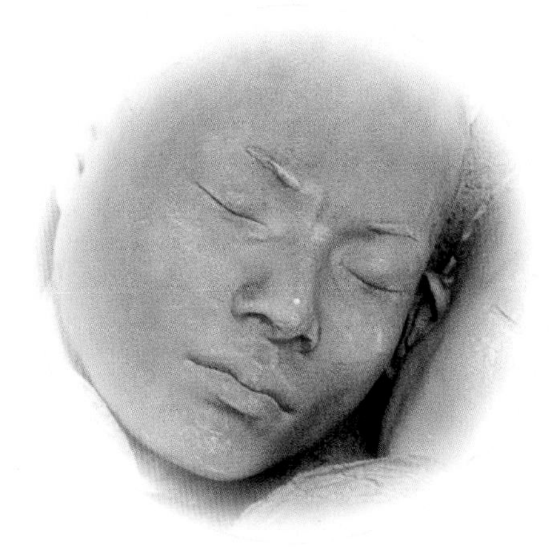

Bird of paradise

17

revealing legs

AMONG THE 750 FULL-PAGE PLATES OF HIS *PHYSICA SACRA*, an elaborate account of every item in natural history either explicitly discussed or even peripherally mentioned in the Bible, Johann Jacob Scheuchzer (1672–1733) surely took most pride in plate 49 of volume 1 (depicted on page 97). For here he illustrated the most fabulous fossil from his own collection, a creature that he had named *Homo diluvii testis*, "the man who witnessed the flood." The great Swiss naturalist exulted in his unique find because he could now provide the hardest kind of objective evidence—real bones—for an event heretofore attested only by the text of scripture:

> We have here not only a figure pressed into stone, and about which one might give full range to the imagination; but we have the very substance of bone incorporated into the rock.

A persistent myth of science holds that raw, hard facts undermine grand ideas and serve as building blocks for more adequate theories derived by patient accumulation. Did not T. H. Huxley, in one of his most memorable aphorisms, speak of "a beautiful theory, killed by a nasty, ugly little fact"? In truth, science rarely progresses in such a simple manner; ideas and information interact in such a complex web that no observation can be fully independent of theory or expectation. Still, undeniable facts can and do intrude upon old prejudices and comfortable certainties—and nothing can be quite so satisfying as this mode of nature's triumph over our mental tyrannies. This essay tells two related stories on this common theme; in each case, a simple, undeniable discovery of something quite humble—a pair of legs—disproved an old reverie and forged a sensible conclusion most helpful to scientific progress.

Exactly one hundred years after Scheuchzer's description, Europe's greatest paleontologist, Georges Cuvier, studied the famous specimen anew. Others had suspected Scheuchzer's error, but none had provided documentation (Petrus Camper, the Dutch anatomist, had proclaimed the flood man a lizard in 1787). Cuvier acknowledged that the large skull matched human dimensions, but expressed incredulity that Scheuchzer could have made such a blunder: "Nothing less than total blindness on the scientific level can explain how a man of Scheuchzer's rank, a man who was a physician and must have seen human skeletons, could embrace such a gross self-deception." Cuvier exposed more bones by chipping away at the rocky matrix. He first excavated teeth of amphibian form. Then, in a crowning proof, he dissected out the front legs (shown in this photograph of the specimen, now in the Teylers Museum in Haarlem), previously buried in the rock—and found "exact correspondence" in form and number with bones of aquatic salamanders. Cuvier concluded: "Thus there is no doubt that the supposed anthropolith [man in stone] . . . was an aquatic salamander of gigantic dimensions."

Most people assume that birds of paradise received their lovely name to honor the exquisite beauty of their elaborate feathers. In fact, they owe their unusual moniker to another legless legend. The natives of New Guinea prepared specimens for trade by cutting off the legs and drying the skin over a fire. The first specimens, supplied by Islamic traders, reached Europe in the sixteenth century and inspired an elaborate legend about ethereal birds that, according to a

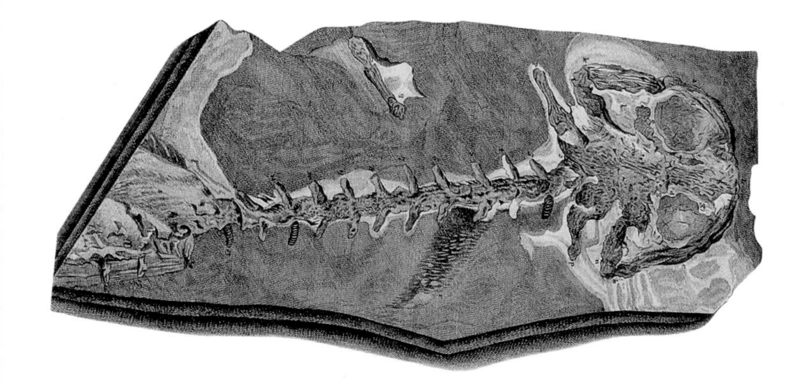

Lithograph of
"Homo diluvii"

Scheuchzer's
"Man of the Flood"

commentator in 1551, "lack feet and are therefore obliged to fly continuously and live in the highest sky. . . . They require no other food or drink than dew from Heaven"—hence, birds of paradise.

Ole Worm (1588–1654), Denmark's greatest naturalist and collector, then proudly debunked this legend by illustrating a specimen from his own collection, with legs prominently attached in the usual place (as in the accompanying photograph, showing a bird of paradise, frayed in feathers, but clearly and eminently footed). Worm made a triumphant appeal to visually obvious factuality: "Those people who believed that these birds lack feet may now be taught the falsity of this view by ocular demonstration."

The most humble bits of anatomy may provide the greatest certainty. We often disparage feet; they may be made of clay in our metaphors; they mock Ozymandias, king of kings, as his only surviving parts. And yet, as Isaiah proclaimed, "How beautiful . . . are the feet of him that bringeth good tidings."

18

pickle of the litter

LIFE BEGAN IN WATER, BUT WE DENIZENS OF THE LAND WON, by consciousness, the right of designation. We therefore, in our terrestrial chauvinism, name many oceanic creatures for a counterpart on terra firma. Oceanic mammals may be sea lions, sea cows, or sea elephants (Seabiscuit, grandson of Man o' War and conqueror of War Admiral, was something else altogether). A casual snorkeler in the shallows may cavort with sea lettuce (an alga), sea robins (fishes), sea hares (snails), and sea mice (worms with the incongruous name *Aphrodita*). But the phylum of echinoderms wins first prize for terrestrial analogues. Three of its five major subgroups have earthy names—sea lilies (Crinoidea), sea urchins (Echinoidea, for urchins, in Europe, refer to hedgehogs, not only to grimy boys), and sea cucumbers (Holothuroidea). Of these three, two rank among the most prominent groups of the fossil record. Sea lilies and sea urchins build skeletons of calcium carbonate, often in small plates numbering several thousand per animal. Some limestones of the American mid-continent consist almost entirely of crinoid stem plates.

Sea cucumbers from the Mazon Creek Formation, shown several times life size (above and following page)

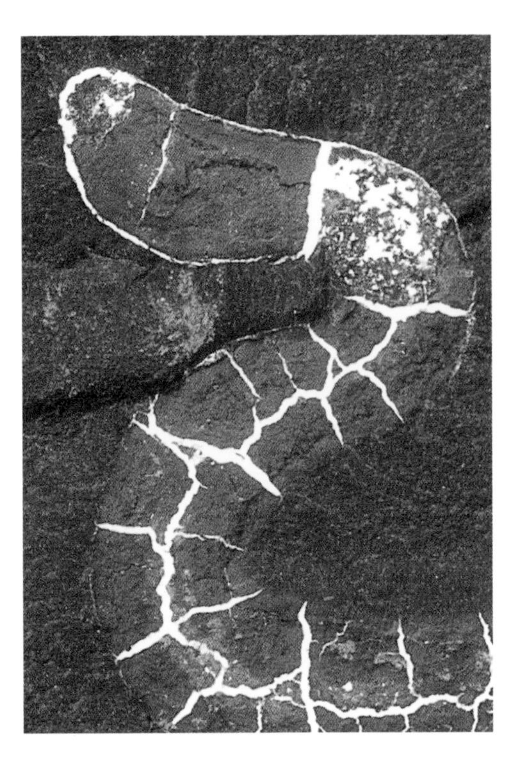

The soft-bodied sea cucumbers, however, rarely enter the fossil record. Some species grow tiny spikes, called spicules, in their body walls, but these microscopic hooks and dots provide no meaningful clues about the form of the body. We must therefore depend upon the most precious of geological rarities—strata that preserve the remains of soft-bodied animals without skeletons—to trace the history of sea cucumbers. Fortunately, each of the three major soft-bodied faunas from the Paleozoic (the first era of multicellular life, ranging from 540 to 250 million years ago) includes a sea cucumber—so we know that this group matches any other echinoderm in geological longevity.

Some sea cucumbers are thick and squat in form, like the vegetable itself, but others grow long, thin, and wormlike bodies. The snake-shaped sea cucumbers in this photograph come from ironstone concretions of the famous Mazon Creek beds of Illinois—our nation's premier contribution to the roster of soft-bodied faunas. Collecting in the Mazon Creek beds includes all the excitement of a crapshoot or a television game show with a car behind one of a hundred panels, and a heap of trash behind each of the other ninety-nine. The fossils lie in the middle of flattened ironstone nodules. You place a nodule on its side, smack it sharply with a geological hammer in just the right place, and the rock splits in two (or the hammer hits your finger, or both). Usually you find nothing; sometimes you recover an incoherent mess, known in the technical jargon of the trade as a "blob." Every once in a while you find an animal—usually an ordinary creature, but sometimes an enigma like the "Tully monster," dignified with the Latin name *Tullimonstrum*, and described by one of my paleontological friends as "a dirigible-like orphan in search of a phylum." Sea cucumbers rank among the rarities—a fit reward after a thousand blanks, twenty-five blobs, and one smashed thumb. These photographs depict the winnowed testimony to concentrated effort, the bright side of a scrap heap.

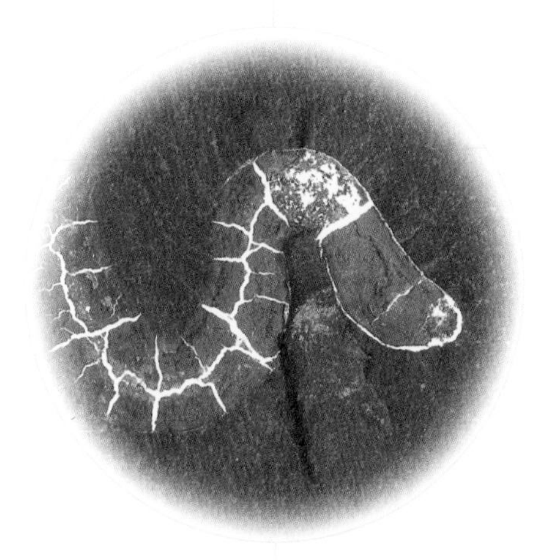

expressing nature's order

Pillar from Watts Towers

19

pride of place

THE MUCH MALIGNED PRACTICE OF TAXONOMY, THE ORDERing and classification of organisms, takes a culturally imposed backseat to the more interventionist and generalizing style of experimentation and quantification in science. But taxonomy should be viewed as one of the most fundamental, and most noble, of scientific pursuits—for what can be more basic than the parsing of nature's rich and confusing complexity? Our categories, moreover, record our modes of thought, and taxonomy therefore teaches us as much about our mental functioning as about nature's variety.

We must taxonomize to make sense of anything diverse and complex. We therefore classify our activities and professions in various ways, just as we order other organisms by Linnaean standards. This essay links two aspects of taxonomy: the literal classification of shells and the ordering of human professions and activities.

Both photographs show geometric arrays of shells in a single species, but the contexts couldn't be more different. Insecure intellectuals make a false, and basically harmful, distinction between "high" and vernacular culture—

and then face enormous trouble in trying to determine a status for significant items in between, like Gershwin's *Porgy and Bess* or the best of popular science writing. These two photographs represent extremes in this false dichotomy. Yet in a deeper sense, both the motives of men and the uses of shells could scarcely be more similar.

The fossil snail shells below, glued to a blue board, identified by a lovely baroque label and photographed upon a marble table made of other fossils, belong to Italy's oldest paleontological museum still intact in its original form—the Museum Diluvianum, established in the 1720s by Giuseppi Monti (1682–1760), professor of natural history at the Institute of Sciences in Bologna. Monti built his "Museum of the Flood" to illustrate the theory, then current, that fossils represented organisms killed and buried in the deluge that spared Noah as guardian for the earth's modern fauna.

The rows of clamshells in the first photograph are cemented into a pillar on the concrete foundation of America's most famous "folk sculpture"—Watts Towers in Los Angeles. Between 1921 and 1954, an uneducated Italian immigrant named Simon Rodia, a plasterer by profession, built his fantastic structure of slender metal towers (up to ninety-nine feet high and made of salvaged steel rods, pipes, and bed

Shells from the Museum of the Flood

frames) on a cement base richly embellished with pieces of glass, tiles, dinner plates, and more than 25,000 shells.

The two major rationales often advanced for dismissive attitudes toward collectors and arrangers may both be dispelled by these twin examples. We do not usually object to unobtrusive and discreet sampling, but blitzkrieg collecting of bushelsful seems more out of place than ever as we approach our shrinking world with enhanced ecological consciousness. Why hundreds of shells (tens of thousands in Rodia's case)? I would hold that, so long as we do not run afoul of limits set by renewability of objects removed from nature, numbers collected must be justified by utility. Shoe boxes and burlap bags full of unusable specimens in museum drawers make me weep, but Monti and Rodia, separated by centuries and maximal disparity of overt intent, amply justify their large appetites by a fully appropriate employment in their own terms: Monti, as he wrote in his catalog, A.M.D.G. *(ad majorem dei gloriam)*, for the greater glory of God, whose earthly bounty includes quantity as well as quality; Rodia for good embellishment of his lifetime's passion. The difference between hoarding and collecting resides in sensible taxonomic order for defensible purposes.

Second, we pour derision on differences that we do not comprehend—the philistine scourge that proper education can reverse. Knowledge may breed both empathy and understanding. Rodia finished his work in 1954 and then moved away from his site forever. The towers, neglected and vandalized, almost tumbled to ruin, but neighbors and scholars saved and renovated this great symbol of blessed obsession, now officially designated a National Historic Landmark. We laugh at Monti's notion that all fossils formed in a single flood, but we forget the historical context that allows us to interpret this view as an advance upon previous concepts of fossils as products of plastic forces in rocks, and not as remains of organisms at all.

The maximally rarefied and the resolutely vernacular; the distant stages of Bologna (site of the Western world's oldest university) in 1720 and modern Los Angeles. Taxonomy's virtue forms the common human bond—so strong an urge to collect, to order, and to use, that we can only speak of a mania almost divine, a spark that may ignite any person in any walk of life at any time, and convert the quotidian to the sublime.

20

brothers under the hair

JACOB DOUBTS THAT REBEKAH'S SCHEME WILL WORK. HOW can her husband, Isaac, now blind and dying, be tricked into blessing her beloved son Jacob rather than his own favorite, Esau? Jacob can imitate his brother's voice, but Isaac will easily discover the ruse: "Esau my brother *is* a hairy man, and I *am* a smooth man. My father peradventure will feel me, and I shall seem to him as a deceiver; and I shall bring a curse upon me, and not a blessing" (Genesis 27:11–12).

In the 1730s the Swiss savant Johann Jacob Scheuchzer conceived and executed the most elaborate popular (and multimedia) work of science ever produced up to then (see essay 17 for more information about Scheuchzer)—the equivalent of Carl Sagan's *Cosmos* television series and tie-in book of our own times. Scheuchzer's *Physica Sacra,* in five volumes with 750 full-page engraved plates, depicts and discusses every event in the Bible with any conceivable implication for natural history. Thus, Moses' second plague inspires a lovely engraving of tadpoles metamorphosing into frogs, which then hop into Pharaoh's bedchamber, while Jesus' designation of evil men as a "generation of vipers" (Matthew 12:34) inspires several plates on the taxonomy of snakes.

Similarly, when Jacob depicts his brother as a hairy man, Scheuchzer produces an engraving of Esau next to the most famous early illustration of a chimpanzee, a mid-seventeenth-century drawing by the Dutch anatomist Tulp—a figure used over and over again in the history of science, most notably by T. H. Huxley in the earliest Darwinian book on human evolution, *Evidence as to Man's Place in Nature* (1863).

Sleeping chimpanzee compared with the Biblical Esau

In our desire to set ourselves apart from, and dominant over, nature—a dangerous position, imbued nonetheless with sufficient biblical sanction, most poetically in Psalm 8—we fear our striking similarity with apes and monkeys. And yet, at the same time, we feel intense fascination with our closest relatives (as a visit to the primate house of any zoo will affirm). This duality of repulsion and attraction leads us to make the obvious comparison—and then to flee from the evident implication of close and meaningful kinship. Scheuchzer himself provides a lovely example of this contradictory pull when he comments on his own engraving of Esau next to Tulp's chimpanzee: "Nonetheless, in making this comparison, I do not wish to insinuate that Esau was a Satyr [then a common designation for chimpanzees], nor that this race of savage animals has descended from him. I consider Esau as a monstrous *man*."

Whatever our fear, whatever our wish to be apart, we cannot escape our affinity with the poor fellow in Samuel Johnson's famous quip about the man who tried so hard to be a philosopher but failed because cheerfulness always broke through. The visually evident evolutionary similarity between ape and human shatters all barriers to an acknowledgment of affinity—no matter how hard we

Endangered gibbon from China

Portrait of Fred Astaire on Ghost Story *set*

try to prevent any breach in our mental wall of desired separation.

Could we possibly suggest a greater difference between any ape and any human than this comparison of the most elegant of all men in the most formal of all attires—Fred Astaire in black tie—with a gibbon, the hairiest of all apes? Astaire, who moved his arms and legs with maximal grace and thrilled us all in film after film—and a gibbon, with hairy arms so long that, in virtually the same pose as Astaire (who folds his hands in his lap), the ape must wrap its arms around equally elongate legs. And yet we look from one to the other, from man to ape, and we feel the visceral certainty of brotherhood.

Consider a final irony. Scheuchzer's Esau and Tulp's chimpanzee mark the minimal distance between an ape and a man (for chimpanzees are our nearest evolutionary relatives and Esau was an especially hairy man). Fred Astaire and the gibbon sit as far apart as an ape and a man can get (for gibbons show greatest departure from us in genealogy among the great apes and Fred Astaire is an exquisitely smooth man). Yet nature breaks through even here, and we know that kinship lurks beneath the disparity of a coat of hair and formal wear.

21

divide and conquer

We usually prefer unification as an intellectual strategy, but in some cases separation and specification of real differences can be even more revealing. "Divide and conquer" does not work only as a cynical motto for expanding imperial armies.

The old parlor game of twenty questions (also good for long car trips) begins by giving the contestants a single hint: the placement of a mystery object among the three traditional kingdoms of nature—animal, vegetable, or mineral. Today we almost "instinctively" parse these three into a dichotomy, with living plants and animals on one side, and inorganic rocks and metals on the other. But previous centuries forged a false unity with several specious arguments, most notably the ordering of all three kingdoms in a single continuum, or "chain of being," rising smoothly from mineral mud to high-minded man (with intentional gender bias). The eighteenth-century Swiss scientist Charles Bonnet, for example, listed asbestos as intermediary between lowly minerals and middling plants (because the stringiness of asbestos crystals reminded him of the internal tubing of vascular plants), and fan corals as the bridge between plants and animals.

Most notably, Linnaeus applied his system of binomial nomenclature—still in universal use for organisms—to minerals as well, thus forging a common taxonomy for all natural objects. This "grand unification" may strike us as silly today, but Linnaeus perceived no anomaly in describing black slate as the species *Shistus niger*, and black walnut trees as the species *Juglans nigra*. We now reject such a standard format because we have learned that rocks and minerals originate by mechanisms so different from the evolutionary construction of organisms that a common division into Linnaean species can only obscure the fundamental disparity.

Species of organisms represent unique historical entities, living in a particular time and place, and generated by a train of ancestors within a genealogical nexus. Each species requires a distinctive name (and Linnaean binomials work well for such a purpose), just as each human being needs a unique designation. *Felis leo* and Frank Leone both provide distinctive names for singular historical objects. And both include a marker of a more general lineage (*Felis* for a genus of cats, Leone for a definite human family), combined with a designation for a particular entity (*F. leo* for the unique population of lions among cats, Frank for an individual male of the Leone family).

Pagoda shells (Thatcheria mirabilis)

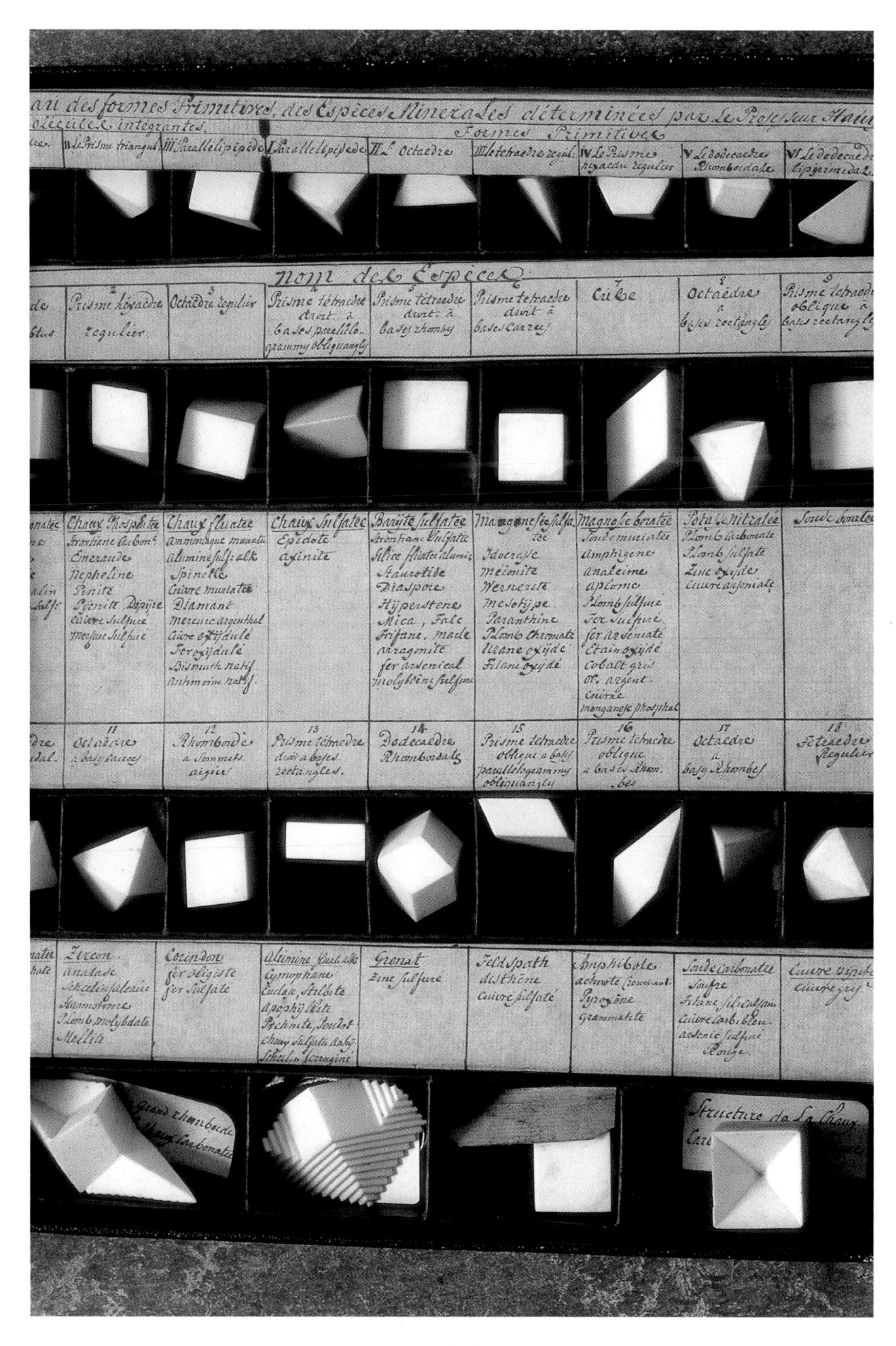

Crystal forms

But minerals are relatively simple, nonhistorical objects built by invariant chemical laws in the same way at all times, and without genealogical connections. A Cambrian quartz crystal (formed 500 million years ago) looks just like a quartz crystal formed last month, because ions of silicon and oxygen join together in a predictable manner. The similarity does not record an unbroken historical lineage of copulating quartz crystals over 500 million years. Thus, minerals should not be named by a system devised to identify unique historical objects built in genealogical continuity.

I view these photographs as a commentary on this fundamental argument about the anatomy of nature. The snail shells, with their sharp angles and geometric regularity, look like objects of the mineral kingdom, built with precise crystal edges by well-aligned molecules. These shells remind me of art deco ornaments for a machine-age architecture (so popular during the 1920s and 1930s), not of the swirling organic forms favored by art nouveau craftsmen at the turn of the century. Nonetheless, these shells are historical products of a particular evolutionary line with a lovely and distinctive Linnaean name, *Thatcheria mirabilis*, but popularly known as the pagoda shell (to honor the building style of another unique historical entity, a human culture this time).

The early-nineteenth-century models of minerals, on the other hand, form part of a collection in the university museum of Utrecht. The upper caption still follows Linnaean principles in designating *espèces minérales* ("mineral species"). But the chosen classification uses the underlying geometry of crystal form—a product of invariant chemical properties for elements building the minerals—rather than any claim for historical linkage among similar shapes. And the caption also credits "le Professeur Haüy"—the great French cleric and scientist, the Abbé René-Just Haüy (1743–1822)—who combined several fields of human study to recognize that he must divide rocks from organisms in order to form a more perfect union in human understanding.

22

stretching to fit

I FORGIVE THE SLIGHT SPIN OF SLOGANEERING CONVEYED by the motto so frequently cited by proponents of a cosmos chock-full of organisms: "Life will find a way." Life *is* resilient and quite capable (especially in bacterial form) of living in the most damnably improbable places—from nearly boiling ponds in Yellowstone National Park to tiny pores in rocks as deep as two miles below the earth's surface. But even this degree of resilience must work within limits; if life ever evolved on the Martian surface during its initial billion years with running water, the planet's later desiccation probably extinguished our solar system's second experiment in advanced carbon chemistry.

The intriguing subject of life's resiliency features two quite separate themes. First, and foremost, naturalists extol the penetration of life into extreme or downright bizarre environments—the hottest, the coldest, the driest, the most noxious pond, the tightest squeeze, the most peculiar and precarious place. But second, and all too rarely considered, we should also emphasize life's internal resiliency—its flexibility to explore the limits of form and structure, not only to inhabit extremes of external environments. To the mite on my eyebrow, the tardigrade encysted for decades of dryness, and the nearly boiling bacterium (exemplars of the first theme of extreme places), I would compare the ant with only one chromosome, the blue whale at an upper limit of size, and the tapeworm at absurd dimensions of twenty feet in length and one-sixteenth inch in thickness (champions of the second theme, stretching the envelopes of possible and workable designs).

Yet just as life can penetrate to most places, but not everywhere (the first theme of resiliency), so too can the anatomies of organisms be tweaked only so far (the second theme of internal flexibility). The realized designs of life do not spread out evenly over the landscape of imaginable form. Rather, islands (often large and crowded but sometimes tiny and isolated) rise into a domain of predominant emptiness. Half a million described species crowd the island of beetles, but think of the vast uninhabited spaces along the trajectory from beetle to bacterium, or beetle to buffalo.

What do these inhabited places represent? Do they specify the designs that cannot work in Darwin's tough world of competition and survival of the fittest—the one-legged mammal that could never outrun the lion? Or do they denote the regions that organisms cannot reach because physical limits upon the structure of matter bar entry —the tree that cannot rise to heaven, or the bird that cannot grow as big as a whale and still stay aloft? Or do they simply mark the domains that organisms have not yet reached because the number of potentially workable designs so greatly exceeds the possibilities for exploration, even given the generosity of geological time? Imagine the conscious octopus that might have evolved (and ruled the world) in a different but sensible replay of life's history.

These three possibilities—the "can't work well" of Darwinian limits; the "can't work at all" of physical constraint; and the "just haven't been there yet" of restricted historical time—bear markedly different implications for our concepts of life and evolution. If the third theme rules, then the order of life's taxonomy primarily records the accidents of a few roads actually taken among myriads potentially available. If the first two themes prevail, then the mechanics of good design and the physics of structural limitation decree a fairly predictable order. I believe that all three themes weave a complex tapestry of rich and conflicting reasons for life's amazing yet ordered variety.

However, one common phenomenon offers a set of best examples for people who wish to deemphasize historical accident, and view life's strikingly discontinuous occupation of potential anatomical space as a predictable combination of physical principles in good design and life's protean capacity to assume, by Darwinian processes, a wide range of form. Life seems able to seize any ecological opportunity, even by altering standard designs to very peculiar purposes.

These photographs present four striking cases of flexibility in this structural sense. All these examples illustrate groups with "standard" and well-known anatomies expressed in so many thousands of species that one might suspect an intrinsic limit to flexibility imposed by key features of design—the mechanical or adaptive impossibility of modifying a key trait, for example. Nonetheless, in each case a few species have evolved radical departures from the standard.

These departures cannot be viewed as either random or capricious. Rather, they bring the oddball creature into a design space usually associated with an entirely different group of organisms—an evolutionary phenomenon called convergence. If a group so firmly associated with one staid form can occasionally transgress into the territory of another, after an extensive journey through the intervening space of organic design, then even the most apparently hidebound anatomies maintain surprising flexibility for opportunistic change—and life will find a way in this internal sense as well.

The first two cases of radical departure present coral- and oyster-like species among the brachiopods. The bivalved brachiopods no longer play a major role in marine ecosystems, but they dominated the early fossil record of animals. Brachiopods persisted as thousands upon thousands of species, usually viewed as pretty "boring" by all but the few paleontologists (including me) who love them—all because they appear to represent so much of a muchness. Most brachiopod species seem to express only limited variations upon a simple theme: two valves, one bigger than the other, and a fleshy stalk, called the

Brachiopod
(Prorichthofenia)

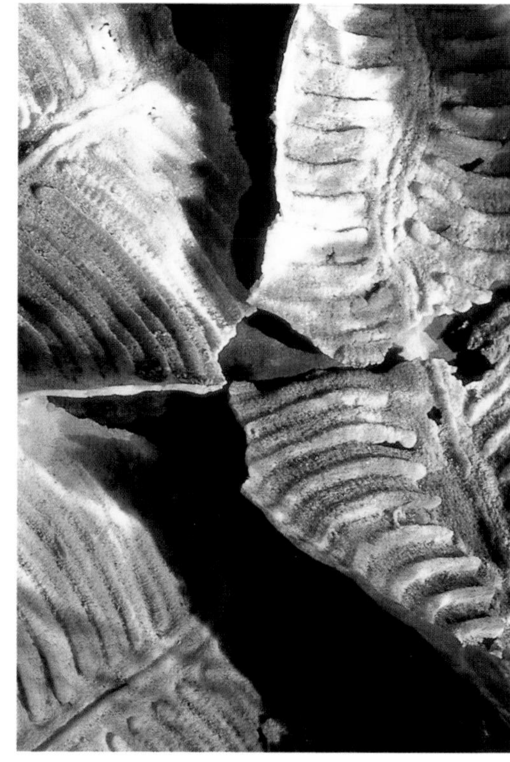

Brachiopod
(Leptodus)

pedicle, that extends through a hole at the top of the larger valve and attaches the animal to the substrate.

Yet, in the two odd Permian species shown in these photographs, both belonging to a group with the promising name of Productacea, these traditions have been abandoned for a remarkable mimicry of the forms and lifestyles of distant taxonomic groups. One species extends one valve into a long cone, reduces the other to a cap—and looks for all the world like a horn coral (and did indeed grow in reef environments). In the photograph on the left we peer directly into the top of the cup, with the cone of the enlarged valve extending invisibly behind.

In the second species one valve becomes irregular, elongated, and cemented to the substrate, while the other grows as a minimal, meandering cover protecting the equally wavy feeding apparatus, called the lophophore, within. These odd brachiopods of the genus *Leptodus* live like cemented oysters and have evolved a striking resemblance to the anomalous form of oysters as well. (Such an "anomalous" design works superbly for cemented creatures that must be flexible enough to grow over and around impediments.)

Fish gotta swim and birds gotta fly. Our third case of departure from the norm violates the usual imperative for birds. The capacity of this large group (represented today by about 8,000 species, most in the category of small aerial tweeters) to evolve flightless terrestrial runners of substantial size has been exploited again and again (see essay 6 for more details and a different take on this subject): the ostrich in Africa, the rhea in South America, the emu and cassowary in Australia, and the extinct moa of New Zealand. In this photograph we see bones of the largest bird of all, the extinct *Aepyornis*, or "elephant bird" of Madagascar—compared with a local species at the upper limits of size for more conventional flying forms, but positively dwarfed by weighty parts of the past.

Our fourth case, the Mesozoic ichthyosaurs, evolved from reptiles of terrestrial design, but became so fishlike on their return to a dis-

Leg bones of the elephant bird (Aepyornis maximus) *with Delalande's coucal*

The flipper of a marine reptile from the era of dinosaurs, evolved from the ordinary finger bones of terrestrial ancestors.

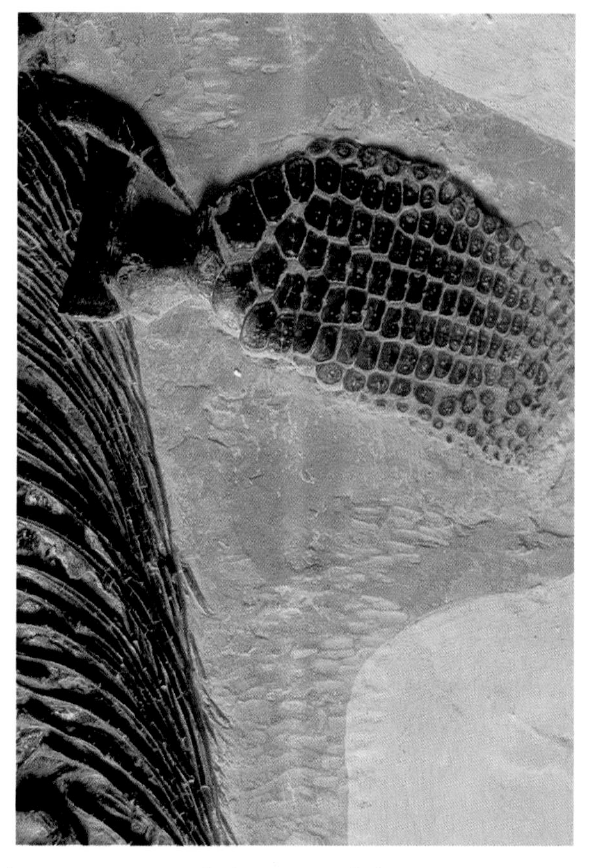

tantly ancestral marine habitat that they even reevolved (from no antecedent structures at all in their terrestrial forebears!) a tail fin with two equal lobes, and a dorsal fin similar in form and placement to the corresponding feature of ordinary fishes. This tail fin has been judged optimal by modern engineers as a stabilizing device for preventing the animal from rolling as it swims forward.

And yet, to end by affirming the complexity of yin's opportunism in combination with yang's constraints, ichthyosaurs did not and could not become fishes again. The imprint of reptilian history could not be completely expunged. But such genealogical baggage may be viewed more as a set of opportunities than an array of brakes. The extended, graceful fins of ichthyosaurs cannot be built, as in modern fishes, of long, slender rays attached to a horizontal base. Ichthyosaurs are stuck with long bones (humerus, femur, and so forth) and fingers. So, instead of growing only three phalanges per finger (as we maintain in all our digits except the thumb, which bears but two), ichthyosaurs evolved fins with twenty or more phalanges per digit. Thus they converted a squat and stubby projection (that, in other vertebrate manifestations, runs on the ground, scratches an enemy, or types this essay) into a long strut, looking for all the world like a fin ray, and permitting a creature from terra firma to find a way back into its own ancestral ocean.

23

animal, vegetable, mineral?

MY ANSWER, FOR THIS HAIRY FISH, MUST INCLUDE ASPECTS of all three—so you get no help at all, by the usual route of restriction for identifying this strange object in only twenty questions.

The basic form may be designated as animal—a fossil fish from the famous lithographic limestone of Solnhofen. But how shall we identify the arborescent growths projecting from all edges of the fossil? They are called dendrites (tree growths) and appear to be vegetable. But they are actually mineral—an inorganic precipitate of pyrolusite, or manganese dioxide. Pyrolusite can form in conventional crystals of tetragonal form, but when carried in solutions percolating through sediments, pyrolusite often precipitates upon bedding surfaces as delicate dendritic growths. Since percolating waters flow preferentially into discontinuities within rocks, and since the fossil fish establishes such a boundary, the mineral dendrites radiate from the animal.

Aspidorhynchus acutirostris, *about three times actual size*

We may, in our sophistication, be tempted to regard the threefold division of kingdoms as a coarse and superficial taxonomy, fit only for an old radio game. But classification represents a powerful strategy for understanding nature, and this tripartite division helped to resolve not only the nature of dendrites but also the basic character of the fossil record. Many scholars of the Renaissance, for example, considered all highly organized objects found in rocks as products of a unitary vital force and therefore properly belonging in a single category. Under such a classification by superficial appearance, scientists could not properly sort out the differing casual bases of complexity—for inorganic precipitates like dendrites, originating within an enclosing rock, could not be separated from remains of animals, deposited into sediments *before* they hardened into stone. When later scientists, following the new Newtonian procedures, built improved classifications based upon genesis and common casual production—rather than similar outward appearance—true fossils finally gained clear distinction from other complex objects, and the emerging science of geology made a great leap forward. "Complex things in rocks" does not constitute a fruitful category for understanding the structures and causes of objects; animal, vegetable, or mineral, though crude, often makes a proper first division.

A final ironic tale about dendrites and the Solnhofen limestones. We usually consider individuality as a mark of *organic* complexity—and regard mineral objects as repeatable (we never speak about Joe or Ida the quartz crystal). But dendrites, with their arborescent, wandering pathways, are complex enough to win individuality—and this property recently came to the aid of Solnhofen's most famous creature: the first bird, *Archaeopteryx*. Sir Fred Hoyle, in his personal retreat to the fringes of science, had branded *Archaeopteryx* as a hoax—a real fossil reptile skeleton with feather impressions squeezed on top into an artificial cement. Dendrites provided a decisive argument against this absurd charge. Many of the feather impressions bear dendrites on top. These dendrites appear both on the slab with the main fossil and on the rock surface (called the counterpart) that pressed against the bedding plane of the main slab (called the part) before quarrymen split the rock to reveal its treasure. Each dendrite is so distinctive that scientists can match the impression on the main slab (overlying the feathers) with the imprint of the same dendrite on the counterpart—"branch" by "branch" and "leaflet" by "leaflet." But if a forger pressed modern chicken feathers into artificial cement, how did the dendrite grow on top of the feathers, and how did an imprint of the very same dendrite also get impressed into the counterpart? No one—not even the best counterfeiter—could have copied such complexity with such exactness on both the main slab and the counterpart. The dendrites, and the underlying feathers as well, obviously existed in the rock before the split that revealed the fossil. The "vegetable" character of this mineral has thus saved the day for Solnhofen's finest animal.

change and directionality

Seven embryos

24

double entendre

WHY DO A LIMITED NUMBER OF GREAT THEMES AND archetypes pervade the sagas of so many different and apparently unrelated peoples? This phenomenon has long fascinated historians and psychologists, and has generally been explained in one of two ways. Perhaps the stories simply spread by learning, and telling, from a single source, and cultures are not nearly so independent as we have assumed. Or perhaps the evolutionary structure of the mind, as shared by all people, encodes archetypes that channel our independently invented stories along similar pathways, no matter what culture invents the tale.

I would add a third reason for similarity: simple logical exhaustion. Some stories can proceed in only a limited number of ways, because the logic of their development (or the limited capacity of our minds) permits no other resolutions.

Our stories about sequential stages seem to follow one of two modes—and I suspect that this restriction falls into my third category of logical exhaustion: we don't know any other way to make such stories "go." We view the stages of our sequences either as increments of progress (simple to complex) or as steps in refinement (ill-formed and inchoate to well-separated and sharply differentiated). The model for the first is addition (each step adds new features and becomes more complex); for the second, differentiation (all bits of complexity exist from the start, but only as potential within an initially homogeneous mass). The "march" from amoeba to human (a false description of evolution) falls into the first category of addition; Michelangelo's assertion that the final statue already exists in the initial block of marble (waiting for liberation by the sculptor) represents the second category of differentiation.

I read this photograph as a depiction of both primal stories at once, as manifested in different aspects of a single sequence of objects. We note a series of embryos preserved in alcohol and housed in a long, rectangular glass jar. Each embryo, suspended by a thread extending from its dimpled head to the top of the jar, casts a wonderfully dappled reflection on the floor in front—where the embryos reappear, head end down, as a series of shadows in a sea of sunlight.

Several of our primal stories can be told in both modes. Most of us read the creation story of Genesis 1 as a tale of addition—first God creates the earth, then plants, fishes, terrestrial beasts, and finally exalted us. But more literal attention to the words strongly suggests differentiation as the intended theme. From an initial formless chaos, God makes a series of progressively finer separations: light from darkness, earth from sky, land from sea, coalescence of sun and moon as sources of light, "bringing forth" of living things from the earth.

Embryology has also featured both stories as anchoring points for the great theories that have defined the subject since its inception. In the major nineteenth-century debate, Haeckel's recapitulation theory —following the model of addition—viewed the embryo as growing ever more complex by repeating the adult stages of ancestors in an evolutionary series. Von Baer's contrary reading—following the model of differentiation—interpreted the embryo as expressing its taxonomic status ever more finely through development: first, one can tell that the creature will become a vertebrate, then a mammal, then a primate, then a human.

I see both stories in one suite of embryos within this remarkable photograph. For me the actual embryos in the jar suggest the model of addition as they increase in size and add visible parts. But the sequence of shadows records differentiation—for all occupy the same length but become more sharply defined with advancing age.

I had always viewed the primal stories of addition and differentiation as our literary biases imposed upon nature's greater richness. But here nature, assisted by art to be sure, seems to be telling us that she acquiesces in these alternative readings of her fundamental sequences.

Human (Homo sapiens), *orangutan* (Pongo pygmaeus),
spider monkey (Ateles paniscus)

25

the yellow leaf road

OEDIPUS PREVAILED BY RECOGNIZING THAT SEQUENCES could be circular, not necessarily linear—time's cycle rather than time's arrow. With his life on the line, Oedipus looked into the mirror of our own life and solved the Sphinx's riddle. He responded by noting that a man rises from the earth but then sinks back down—moving on four legs as a crawling child, on two as a proud adult, and on three in great age, stooped and supported by a cane.

Somewhere along the way of Western history, as progress became a dominant metaphor in the seventeenth century and later, we lost this insight about the potential complexity of temporal sequences. Although evolution, as scientists understand the process, proceeds by nondirectional branching and meandering, our cultural biases lead us to depict the history of life as a progressive march to better and higher states. What can be more familiar than the canonical icon of a stooped monkey leading to a half-erect ape and culminating in an upright human (usually, in another and more immediate bias, depicted as a white male in a business suit)?

Mangabey (Cebus apella) *with wave-washed book*

This modern parody of our most constraining bias about nature recovers Oedipus's insight. Purcell has aligned three primate skeletons in reverse order—the ape following the monkey, and the human last in line. These specimens, from collections of the Universiteitsmuseum of Utrecht in the Netherlands, have been moved outdoors, to the courtyard of the university library, and set upon a yellow leaf road. (Purcell took these photographs in late fall and imported a bag of yellow leaves from a nearby ginkgo tree to construct the preferred pathway. The leaves bear modern witness to an interesting historical episode, when Japan, as a closed nation, permitted only a few Dutch ships to land each year for their totality of foreign contact. The tree, now more than two centuries old, grew from a Japanese seedling imported during this period of minimal cross-fertilization.)

The leading monkey, knees bent, back curved, and arms extended, heads the parade in conventional posture. The ape, an orangutan with less stoop and curvature, holds the usual middle ground. But the human being, the skeleton of a man with rickets (a vitamin D deficiency that can cause severe spinal curvature), sports the conventional symbols of "primitive" monkeydom in even more exaggerated form—legs even more bent and spine even more curved. He trudges slowly along, last by default, I suppose, aided by the "walking stick" that supports his skeleton in the museum's collection. The solution of Oedipus has been fulfilled—and why not? The yellow brick road eventually leads from Oz to Kansas, not from primitivity to Parnassus.

While we depict other primates as primitive, we also sense—with a discomfort that we often cover by humor—their close affinity with our exalted selves. Thus, in another conventional icon, we attempt to reassure ourselves by placing monkeys in distinctive human situations, and laughing at the incongruity (or are we really depicting the uncomfortable inverse and mocking our pretensions?). Haven't we all seen the statue of an ape, posed as Rodin's thinker and either contemplating a human skull or reading a copy of Darwin's *Origin of Species?*

In the second photograph a mangabey monkey, tail bent upward to permit a sitting posture, rests on a picnic table covered with green oilcloth and a late-autumn windfall of cherries. In its hand, Purcell has placed a little book, found on a Dutch beach and given to the Utrecht collection by a man engaged in studying the action of waves and water upon objects. I could only think of Alexander Pope's heroic couplets on the inferiority of human intellectual preeminence as seen by the putative higher creatures of other planets. Our parodies puncture our prejudices:

> *Superior beings when of late they saw*
> *A mortal man unfold all nature's law*
> *Admired such wisdom in an earthly shape*
> *And showed a Newton as we show an ape.*

Collection of rhinoceros horns

26

preposterous

IN A NUMBER OF VERSIFICATIONS OF BOOKS FOR CHILDREN, rhinoceroses receive a one-word definition in a near rhyme: preposterous. The five living species of rhinoceroses, viewed as tanklike vestiges of a prehistoric past, and barely hanging on as threatened populations in their African and Asian homes, do convey an image of superannuated heavyweights from a lost world where brawn could make up for stupidity and ensure survival.

Modern rhinoceroses do represent a remnant of past glory: These animals were once maximally prosperous, rather than preposterous. Their enormously successful fossil forebears included *Paraceratherium*, the largest land mammal of all time—eighteen feet high at the shoulder and a browser of treetops. Their extensive ecological range included small and lithe running forms no bigger than a goat (the hyracodontines), and rotund river dwellers that looked like hippopotamuses (the teleoceratines). Moreover, modern rhinoceroses represent a vestige within a vestige. The formerly dominant order of odd-toed hoofed mammals has now dwindled to three groups: the rhinoceroses and the tapirs, each barely hanging on, and the horses, given an artificial boost and a new lease on life by human needs for transport and human foibles for warfare and wagering.

A dilemma, in technical terms, is a problem with two logical solutions, each untenable or unpleasant. We speak of being caught on the "horns of a dilemma," in reference, I suppose, to the crescent moon with its two points, or horns (or perhaps to the devil himself). The dilemma of the rhinoceroses also rests upon two aspects of their distinctive and defining horns. On one point, horns mark the rhinoceros's fascination as both preposterous and alluring—a sign of fame and therefore a desired trophy for Western hunters. On the other point, horns inspire legends of utility for alleviating various human ills, particularly sexual impotence in males—and the few remaining horns have therefore become a prize for poachers and a substance nearly beyond price in Eastern pharmaceuticals.

We all experience a personal breaking point, where moral indignation swamps dispassionate analysis. I can read about human desecration of animals with reasonable equanimity in the face of great sadness, but I fall into predominant anger when I encounter the numerous stories of magnificent creatures slaughtered in vast numbers for single parts deemed desirable (often so frivolously) or useful (often so fallaciously) to humans: elephants only for their tusks, buffaloes for their tongues, nightingales for the same organ (considered a prime delicacy in Roman banquets), and gorillas because some people like to buy ashtrays made of large primate paws. Rhinoceroses, sadly, become victims of this same outrage: their most distinctive evolutionary markers serve as the brands of their destruction.

I wish I could portray naturalists as perennial opponents of such exploitation. We certainly function in this manner today, but our past does not always measure up to current practices. In a former age, that viewed nature as an unlimited bounty and men as masters of all, naturalists often collected in a wanton manner—as though scientific study demanded mass transport from live in the field to dead in a museum drawer. This photograph depicts just some of the specimens in a "miscellaneous bin" of rhinoceros horns in the collection of my own institution, the Museum of Comparative Zoology at Harvard.

These specimens range from horns bought by the museum director Alexander Agassiz from Ward's of London in 1877 to animals shot by hunters on safari and donated to the museum as late as 1936. Detached horns, often severed with pieces of surrounding skin. The part that dooms the whole. A strangely beautiful picture of elegance separated from a symbol of ungainliness. Do we not witness here the moral equivalent of preposterous?

27

less stately mansions

DIVERSITY MUST BE HONORED AS NATURE'S IRREDUCIBLE signature. We often avoid this primal fact by choosing one specimen as a "type"; the rest then pale by comparison, or simply pass from consciousness, as "standard" illustrations clone the chosen representative. Often the conventional standard badly misrepresents the group, either because the "type" specimen stands at one extreme in a range of diversity or simply because the group's variety ranges too widely for any single object to capture. Dick and Jane, after all, did not honor our full diversity as embodiments of *Homo sapiens*, juvenile version.

We think of opossums as a primitive remnant of early marsupial mammals, the conventional "living fossils" of our textbooks, because we focus on the so-called Virginia opossum as a type. But opossums include a vigorous group of some thirty species, actively evolving in South America. Our type represents the only member that invaded North America after the Isthmus of Panama rose on a geological yesterday. We think of rhinoceroses as slow, lumbering, and primitive—altogether antediluvian. But this diminished group of

Early Paleozoic straight-shelled nautiloids (Orthocones). Note the single slightly coiled relative. They are all cousins of the modern highly and fully coiled chambered nautilus.

mammals once included a branch of sleek runners (the hyrachiids), and a short-legged, hippolike division (see previous essay).

The chambered cephalopods known as nautiloids (relatives of octopuses and squids) also receive short shrift in our consciousness because their past glory has been reduced to a sole survivor markedly different in form from its progenitors. This survivor—the chambered nautilus, Oliver Wendell Holmes's ship of pearl—does not set a shabby standard, to be sure. This species has become our primary symbol of nature's fragile beauty, even a potential guide to our moral development—"build thee more stately mansions, O my soul," as Holmes wrote in comparing our ethical growth through life to the shell's increase in size.

But the chambered nautilus is only a remnant of past diversity, and a skewed remnant at that. The nautiloids evolved as nature's first great group of predatory animals, the top carnivores of the sea for a hundred million years before the rise of fishes. They then held their own (with help from their closest cousins, the ammonites) until the same extinction that killed the dinosaurs also wiped out ammonites some 65 million years ago. The nautiloids persevered through this mass dying, but only as a shadow of their Paleozoic variety.

In their reigning days, most nautiloids grew in a shape quite contrary to the modern type—namely, as straight tubes, the famous chambers stacked one atop the other. I love the lonely ship of pearl, but I also marvel at the vigor of these diverse ancestors, some 450 million years old. The fossils in these photographs came from the famous collection of a nineteenth-century Bohemian beer baron. He cut and polished many of the straight shells to reveal the chambers and the connecting tube among them; he wrote his identifications right on the specimens in a broad and graceful antique hand.

I also have good news for fans of modern nautiloids. The chambered nautilus may not be so lonely as we once thought. Recent genetic and biogeographic studies indicate that several distinct species can be recognized under the mask of a shell that varies little in form. Nautilus may represent only a shadow of its past, but nature's motor of diversity continues to run.

28

consider the lilies of the sea

IN THE PHOTOGRAPHS OF CRINOIDS THAT FOLLOW, WE stare across the greatest divide in paleontology. Brief episodes of mass extinction have punctuated the history of life and established many primary patterns in the pageant of evolutionary change. (Mass extinctions do not merely accelerate or retard a set of more stately changes working their inexorable way through the normal times in between.) The most momentous, by far, of all mass extinctions occurred 250 million years ago. This debacle ended the Paleozoic Era of our geological time scale and may have wiped out more than 95 percent of marine invertebrate species.

Echinoderms may not rank high on the hit parade of invertebrate favorites, but we should at least develop a parochial regard for their welfare, since they rank as our closest genealogical cousins among the invertebrate phyla. Of the five modern groups of echinoderms, crinoids rank as the least diverse and least noted (compared with sea urchins, starfishes, brittle stars,

Silurian crinoids before the great Permian extinction

A Carboniferous crinoid, also before the Permian extinction

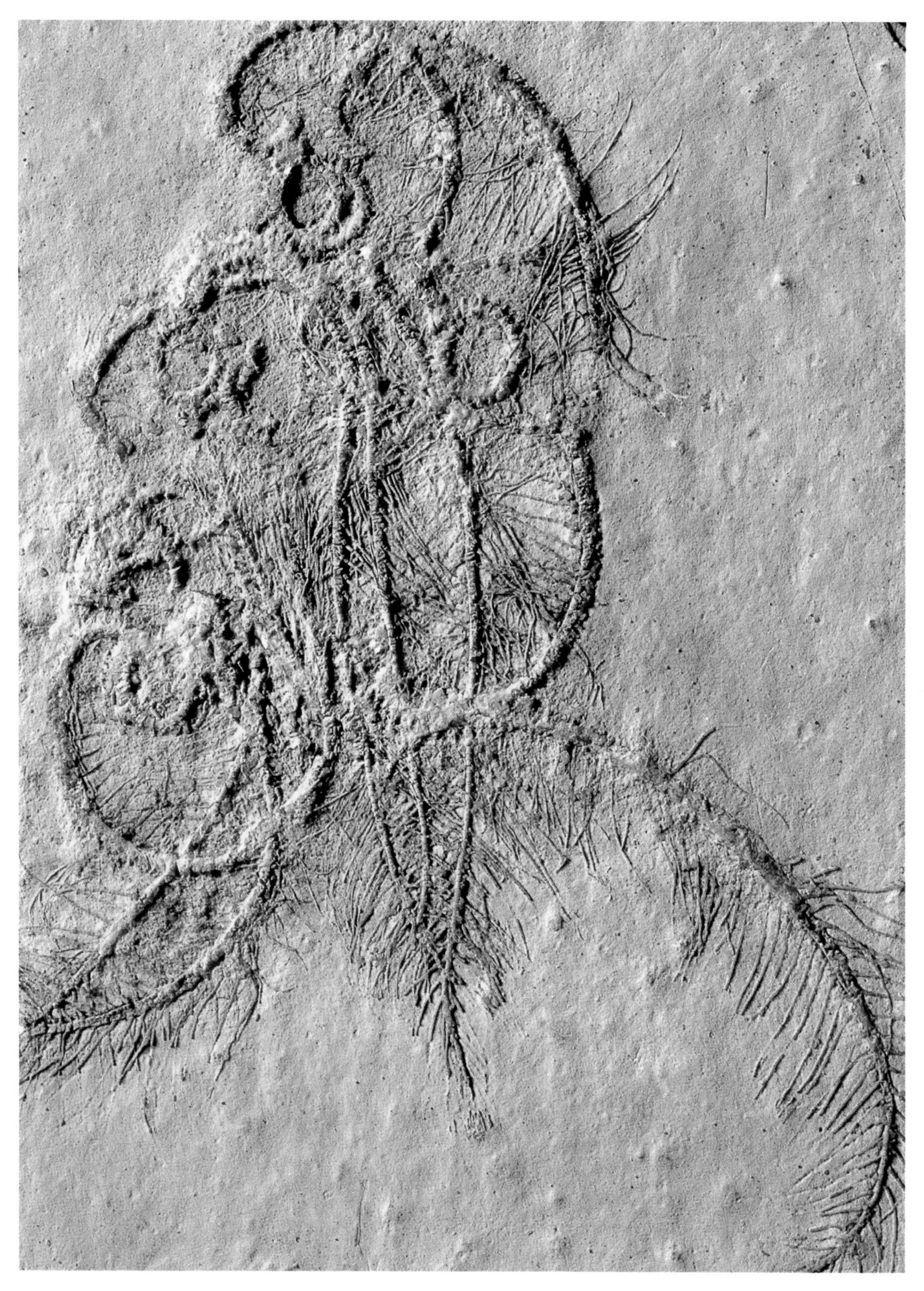

Saccoma, *a delicate stemless crinoid, evolved from a lineage that survived the Permian extinction*

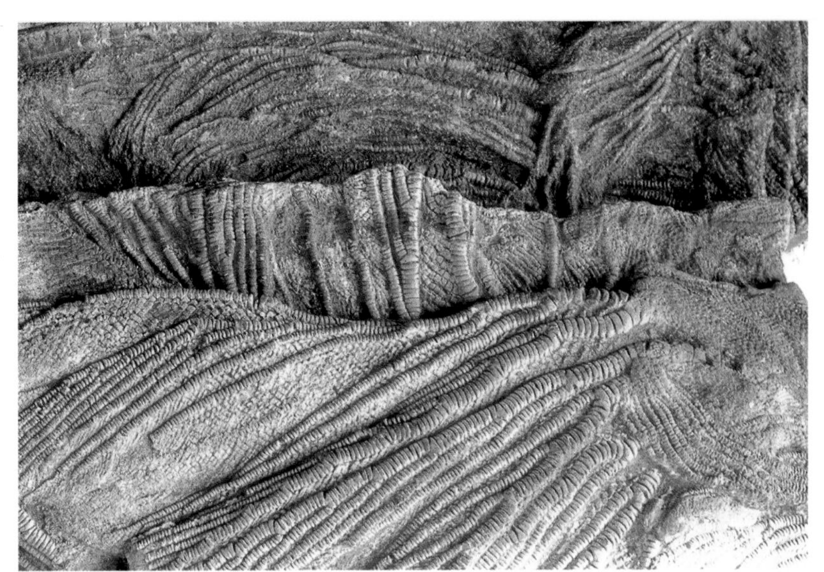

Arms of Pentacrinus, a lineage that survived the Permian extinction

and sea cucumbers). Yet crinoids held court as the most conspicuous of echinoderms before the great extinctions permanently whittled their diversity to a fraction of its former glory.

These four photos present a set of before and after mug shots in the history of crinoids and their fortune across the great divide of the Permian extinction. Echinoderm skeletons include large numbers, often thousands, of individual calcite plates. After death, the animal usually falls apart into as many pieces, and most crinoid fossils end up as disarticulated, and uninterpretable, hashes. But, frequently enough, circumstances conspire to preserve the animal intact. A "standard" crinoid includes three basic parts: a stem (with a root system, called a holdfast, at the base) built as a single stack of round or polygonal calcite plates; a cup (or calyx) mounted atop the stem and composed of calcite plates surrounding the main body of the animal; and a set of arms, extending upward from the cup, and used primarily in feeding.

Paleozoic crinoids (before the extinction) tended to build large rigid cups and strong stems. The Silurian slab shows two specimens, with upper parts (cups and arms) facing each other and stems pointing away to upper left and lower right. The sturdily built Carboniferous crinoid shows all three divisions of arms, cup, and stem against a gray limestone background.

Mesozoic and later crinoids (after the extinction) construct smaller cups and longer, highly flexible arms. This trend to greater mobility reaches its greatest extent in a group (including most living crinoids) that dispense with their stems altogether, thus gaining the ability to crawl, or even to swim actively, from one feeding site to another. *Pentacrinus* denotes the classic genus of long-armed crinoids, and this specimen includes little more than a forest of extended arms. *Saccoma*, from the celebrated Solnhofen limestone (site of *Archaeopteryx*, the first bird, and other beautiful soft-bodied fossils), is a stemless form with long, delicate, and wavy arms—perhaps a swimmer.

These trends—from massive, sturdy, inflexible early forms with their rigid cups to delicate modern forms with small cups, no stems, and long, wavy arms—often inspire an interpretation of progress in biomechanical excellence. A leading modern textbook, for example, states: "The acquisition of brachial [arm] flexibility and liberation from the stem were undoubtedly of great importance in determining the evolutionary potential of this crinoid group."

But I demur. The facts of diversity in the fossil record speak differently—and the assumption of improvement only records our disposition to grant the history of life a heroic character of progressive advance. In fact, crinoids were decimated at the great extinction and never recovered even a fraction of their former abundance. Paleozoic crinoids fall into three great genealogical groups; all later forms belong to one surviving lineage. Paleozoic crinoids built "forests" in shallow seabeds, and great thicknesses of limestones consist almost entirely of their disarticulated remains; later crinoids scarcely impose themselves on our consciousness at all. *Sic transit gloria mundi*.

Malmaison I (construction)

29

infiltration

WE ALL KNOW, FROM THE MOST PEDESTRIAN EXPERIENCES of our daily lives, that order (both mental and material) can only be maintained by ceaseless struggle against a pervasive, universal push toward completely undifferentiated chaos. This tendency toward collapse into a formless, mixed-up glop affects all systems at all scales, from the impetus for the most particular parental admonition—"Stevie, clean up your room or I won't let you go out and play stickball"—to the grandest and most general physical principle: the second law of thermodynamics, with its proclamation that, in closed systems, entropy (a measure of disorder) must continually increase. This natural infiltration of chaos into order also serves as a metaphor for the adulteration and debasement of human decency, as our bastions of constructed beauty and kindness corrode before the besieging forces of war and bestiality.

Nineteenth-century science formulated the principle of entropy, but European faith in progress and rationality prevented this truth from seeping into popular consciousness until the most cataclysmic event of our century opened the floodgates of despair. I refer not to the Second World War, or the Holocaust, or Hiroshima, for the old faiths had been irrevocably shattered by then. Rather, the turning point must be located in the senseless carnage of the First World War, the greatest breaker of illusions in Western history.

If windows represent our standard image for communion between our subjective souls within and an objective nature without—the glass through which we see reality but darkly—then this tetraptych presents a hauntingly complex and multilayered picture of infiltration in both domains. This window depicts the natural infusion of disorder through universal principles of entropy, and the creeping influx of war and bestiality into the eternal human hope that art and virtue might stem the tide both of our own failings and of nature's ways.

The substrate and structure for all four panels—the chair in the lower-left panel, the rugs and the tiles, the picture frames in the lower panels and the display cabinets and bathtub above—depict objects from Malmaison, the house of Napoleon and Joséphine, where the former empress lived until her death.

But in Purcell's work this island of artificial order has been infiltrated by the human institutions (war) and human symbolisms from nature (bestiality) that signify advancing chaos. The human inputs all record the carnage of the First World War. On the staircase in the upper-left panel, an old rotogravure shows an army officer (we see only his hand gripping the lower part of the banister) about to arrest a "traitor" (the body at the top of the stairs, now capped by a skull from a colonial New England gravestone). The same panel shows the spoils and consequences of war—a wheel from a war cart left to rot on a battlefield, and a winged statue from a Liverpool war memorial just below. In the lower-left panel, bombed-out churches form the backdrop to Napoleon's seat of former and temporary power.

The frames and furnishings that once held signs or forms of elegance have been infiltrated with symbols of bestiality—the sharp-beaked bird and the New Guinea "savage" in the lower-right panel (representing human animality in the racist beliefs of those who had just perpetuated the true savagery of the First World War); and above, in the cases that once held porcelain or silver plate, the head of a condor, a raptor that feeds on putrid flesh. Meanwhile, the ultimate symbol of the beast within us, a monkey with sharp teeth, cavorts in Napoleon's own bathtub.

Nature may bat last in some ultimate, thermodynamic sense, but our own unique biological gift, the inherent flexibility that can build order by using both art and decency, could also forge a mighty fortress to postpone this Waterloo into a comfortably distant future—if only we could find the will and the way to suppress the dark capacities on the other side of this same blessed flexibility.

30

from man to mouse

Nothing of him that doth fade

But doth suffer a sea-change

Into something rich and strange.

In this eerie ditty from *The Tempest*, Shakespeare describes the transformation of human bones and eyes, lying "full fathom five" at the bottom of the sea, into coral and pearls. The photograph here cannot match the Bard's imagery for magic and mystery, but at least we observe a "land change" into something useful, if only for some rather distant evolutionary relatives. In a New England junkyard, Rosamond Purcell found a stack of books piled onto a board and left outdoors to a natural fate. The treasure included two books that had fused together on the ground at the bottom of the pile. The lower book remains relatively intact, but the left half of the upper book has been reconstructed by local mice—chewed to bits, gouged out, and stuffed with straw to build a nest. Purcell could just make out the former name of the current nest, a romantic adventure entitled *Flying Hostesses of the Air*. Thus have the local mice converted metaphorical into literal pulp.

One should not wax overly philosophical about a humble and concrete object of such evident utility, but this book-nest does serve as a striking symbol for a central issue in the highfalutin realms of both ontology (the nature of being) and epistemology (the basis for our knowledge of being). In a universe of constant flux and transformation, how can we characterize the evident (if transient) stability of objects, and how can we recognize and name such objects if they gradually change into something else over time? In other and more concrete words, how and when does a book become a nest?

The problem remains minimal for sequences based on simple changes in geometric proportions, with no major alteration in function—the evolution, say, of a wolf into a dog. We do face the classic "naming dilemma": when does one become the other, and how shall we designate creatures living in the middle of such a smooth and slow transition? Nonetheless, we continue to recognize both wolves and dogs as basically the same kind of thing.

But how shall we treat a smooth transition between two objects that ordinary language and common sense yearn to identify as members of basically different categories—in this case, a readable human book and a nestable murine home? I cannot solve the general problem (nor can anyone else, for that matter), but I can describe a humble guide, often used by students of biological evolution, to clarify the question. In my trade, we treat this issue by trying to separate two components that generally run in different courses during a transition: structure and function.

Functional change will generally be more rapid, more complete, and more irrevocable—thus inspiring our sharply separate names for the end points. In functional terms, the book can no longer be read and has therefore ceased to exist; the object now works as a nest for a different species.

But structural change will rarely erase every vestige of past existence, even while repackaging the original substance for new purposes. In structural terms, we can still know that this nest passed through a former life as a pair of books. The bottom book remains nearly intact, even though the pages have fused to unreadability. The top book may be shredded beyond any possible recovery of a flight attendant's tale, but we can still read a few words on some scraps of nesting. Interestingly, I can make out only five complete words, all contributing to the moral of this tale: "to use," a phrase that may symbolize the process of transformation to different ends; "metal," a mark of inherent structural worth (our English word *mettle* has the same etymological root); and, on a third scrap, both "and" and "the"—our two most common words, and therefore the most likely structural survivors of such a thorough functional transformation.

So long as the wheel of time's fortune obeys this frugal and eminently "green" strategy of recycling for sustainable use, the universe may never run down. If a phoenix can arise from ashes, why should a book not nurture the next generation of mice?

The Book Nest

afterword

WHEN THE SUBJECTS I SELECT OR THE CONSTRUCTIONS I have built are reduced to the two dimensions of a photograph, they may gain in emotional complexity. I am drawn to the literary and historical associations that emanate from objects rather than to single definitions of them, and am attracted to metamorphoses achieved either by nature or by design. For example, one of my most cherished possessions is the book that has been turned into a nest of syllables by a mouse or squirrel. Rather than photograph a game of chess, I construct a game board of subliminal players and photograph the construction. I work with natural light in museum collections or in the studio. In this fashion, I photographed the two-headed sheep in golden liquid, a conjoined headed child, anatomical wax figures more glisteningly real than the real, a tapeworm looking like a terracotta model of a maze, a lightbulb with what appears to be a Japanese landscape etched by fire, and a tin toy split by age set in a patch of refracted sunlight. The stuffed gibbon *(Hylobates concolor)* and Fred Astaire *(Homo sapiens)* project the same attentive gravity—the gibbon, the self-possessed aristocrat of the Chinese mountain forests, now highly endangered, and Fred Astaire, the self-possessed star, no longer dancing, trapped here in moody advanced age.

Taken as a group, the contents of these pictures represent, of course, a kind of virtual cabinet of well-worn curiosities. Many of the artifacts come from one esteemed museum or another—the ammonites from the University Museum at Oxford and the Sedgwick Museum in Cambridge, England; the elephant bird egg from Madagascar by way of Leiden's Naturalis Museum; and the human anomalies from the Warren Museum in Boston and the museum of Cesare Taruffi at the Instituto di Anatomia Abnormale in Bologna. By associating disparate artifacts across time and space, we suggest the seeds for a collection of ideas rather than a physical entity under a single roof. As Stephen Gould reflects on evolution, entropy, and the nature of things, I study both natural and cultural artifacts for visual signs of life, of change, of transformation. The pairs of photographs—for many of these essays include two pictures—may seem, at first glance, unlikely partners. Together, they send down roots. For me, the baroque museum collection of shells glued to dark blue paper and the bivalves embedded on the pillar's capitol of Rodia's Watts Towers form a kind of part/counterpart of the human desire for order—that of the methodical collector of natural history and that of the devoted builder of an ideal structure. Rodia's creation seems more like a ship driven inland than a tower. Taken together, the two collections of shells provoke images of ships, the Flood, Noah's Ark. Although I am neither sculptor nor architect, I have made a number of constructions out of found elements (see page 148). This work is small—nothing as grand as the Watts Towers—but because it often defies a single definition, I feel an affinity for Rodia's style. As an artist, my methods are far from the taxonomic music made by Monti's shells on cards. While I'm still mentally cresting the waves of the original flood, however, Gould, in his essay on the shells found continents and centuries apart, refers only in passing to Old Testament events before he moves decisively into the world of the history of science and of human motivation.

Everyone sees what they want to see and what they are used to seeing. I do not expect unilateral agreement from viewers about what these pictures "mean." I prefer to show the pictures rather than talk too much about them. I will describe the contents and intentions behind a photograph to demystify its superficial strangeness: "The skull from the collection of Joseph Mutter came from the catacombs somewhere in eastern Europe; it lies on a chromolithograph by the master draughtsman Jacques-Fabien Gautier-Dagoty of the interior of a woman's body. The artifacts come from the Mutter Museum, College of Physicians, Philadelphia, PA." These are concrete facts. The bringing together of the skull and the womb evoke the cycle of life, fertility, and death, of life in death, and of the brevity of human existence. (Now, aren't you glad you asked?) Of course, you may not see it the same way. Gould may not see it the same way, either. That's okay by me. While I appreciate anyone who takes the time to see past a signifying object to the ideas beyond, my primary concern is to produce the visual evidence.

As scientific evidence the photographs become half of a new metaphoric equation. In this book (where so much comes from gently eccentric sources), a scientific yardstick applied to aesthetic effect yields, we believe, a richer mix. No bird of paradise has ever lived near a giant salamander, but bird and reptile share a "footless" reputation; the baroque dentist never extracted a mastodon tooth, but teeth mean a great deal to both humans and pachyderms. Gould circumvents superficial nevers, bringing together organic, historic, and allegorical arguments, telling us more about birds, animals, dentists, and elephants than a picture ever could. He crosses back and forth between the territories of science and human history, constantly weaving together the "sites of broken knowledge" supplied by the photographs. He makes them work (as Krystof Pomian has described seventeenth-century objects) as semaphores, signifying not just the physical reality or appearance of an object, but a wealth of underlying meanings—natural, cultural, and visual. From his explanations, I often experience a frisson of insight into a scene I had myself recorded and therefore thought I knew.

I thank all the collection managers and curators who helped me photograph the material. We thank especially the three people who have shown such faith in our work—Christopher Reed *(Harvard Magazine)* and Peter Brown and Elizabeth Meryman *(The Sciences)*.

ROSAMOND WOLFF PURCELL

THE PHOTOGRAPHS HEREIN PREVIOUSLY APPEARED IN:

Special Cases, Natural Anomalies and Historical Monsters
(San Francisco: Chronicle Books, 1997):
pages 22, 26, 29 (top), 52, 109, 136

Curiositeit (Utrecht: Universiteitsmuseum, 1997):
pages 45 (top), 59, 60, 91, 115, 132, 134

Finders, Keepers (New York: Norton, 1992):
pages 29 (bottom), 31, 36, 68, 78

Suspended Animation: Six Essays on the Preservation of Bodily Parts
(New York: Harcourt Brace, 1995): pages 65, 128

Swift as a Shadow: Extinct and Endangered Animals
(Leiden: Naturalis Museum, 1999): pages 45 (bottom),
110, 111, 121

about the authors

STEPHEN JAY GOULD is the Alexander Agassiz Professor of zoology and professor of geology at Harvard. He lives in Boston, Massachusetts, and New York City.

ROSAMOND WOLFF PURCELL is an artist and photographer who lives in Boston.